AF556847

NOVEL NDE
Methods for Materials

NOVEL NDE Methods for Materials

Proceedings of a symposium sponsored by the Chemistry and Physics of Metals Committee of The Metallurgical Society of AIME and the Electrical, Magnetic, and Optical Phenomena and the Structures Activities Committees of the American Society for Metals held at the AIME Annual Meeting, Dallas, Texas, February 15-17, 1982.

Edited by
BHAKTA B. RATH
Material Science and Technology Division
Naval Research Laboratory
Washington, D.C. 20375

A Publication of 

The Metallurgical Society of AIME

A Publication of The Metallurgical Society of AIME
420 Commonwealth Drive
Warrendale, Pennsylvania 15086
(412) 776-9000

Printed in the United States of America.
Library of Congress Card Catalogue Number 83-62891
ISBN Number 0-89520-466-5

FOREWORD

Because of an acute need for increasing national productivity, material durability and reliability, and establishing retirement-for-cause criteria, the importance of nondestructive evaluation (NDE) of structural materials has steadily increased in recent years at an unprecedented rate. Recognition of this need may be illustrated using the gas-turbine engine components where 99% of the components are prematurely rejected because of our lack of knowledge in accurately determining their useful life. This is only one of numerous cases where better life prediction through NDE is essential. Furthermore, an increased emphasis on NDE in recent years is also in part due to demands on enhanced performance of structures and systems and development of new classes of materials using novel processing and fabrication technologies. The new materials, such as powder consolidates, composites, and ceramics require new and reliable detection methods for evaluation of fine flaws and defects, which has thus far not been possible by the conventional NDE methods.

The major thrust of this symposium was to highlight behavior of materials as well as response of innovative techniques which have excellent potential for development of new NDE methods and procedures. Recent studies have indeed demonstrated the use of long-wavelength neutrons, positrons, x-rays, lasers, Rayleigh waves, and vibrothermography for nondestructive evaluation of flaws and defects in solids. The papers presented in this symposium will provide a single source for scientists and engineers on further exploration and application of new methodology to NDE.

The Editor wishes to acknowledge the assistance of Prof. J. T. Waber of Northwestern University and Dr. P. Parrish of the Army Research Office for their assistance in the organization of this symposium. The active participation of Dr. L. A. Jacobson of DARPA , Materials Science Division, Dr. S. Spooner of the Oak Ridge National Laboratory, and Prof. R . E. Green, Jr. of Johns Hopkins University is greatly appreciated.

Bhakta B. Rath
Material Science and Technology Division
Naval Reseatch Laboratory
Washington, D.C. 20375

August 1983

CONTENTS

POSITRON ANNIHILATION - A NON-DESTRUCTIVE PROBE

FOR MATERIALS SCIENCE

James T. Waber

Department of Materials Science and Engineering
Northwestern University
Evanston, Illinois 60201
USA

A number of topics of interest to a material scientist will be covered in this review. The traditional non-destructive use of positrons is to study various kinds of defects in the lattices of metals and ionic crystals. Problems arise when one tries to urge a specific interpretation on the results and in general a consensus has not been obtained. There are effects which do not correlate well with annihilation at defects in the bulk of a solid. The experiments in many cases however, were not done carefully enough to exclude the strong influence of the free surface of the solid as well as the contamination on internal "surfaces" such as voids and/or grain boundaries. Specimens in the past were not always well characterized. Surface effects may have plagued interpretation.

This effect of the annihilation near the surface may indeed be an important advantage of using positrons. The positron has a negative work functions and thermalized ones tend to drift to a free surface. Apparently they reside there for a short time before annihilating. If that is the case, they could be a powerful probe of the condition of the surface and of any species adsorbed on it.

In the case of either "high energy" electrons or positrons, they penetate much deeper than the top surface layers. In principle, the positrons do not have one advantage that the electrons do for surface analysis, - the electron which can be detected outside emerge from a shallow layer, roughly 1 to 10

Work supported in part by U. S. Army Research Office.

nanometers thick. The energy associated with adsorption or disappearance of the bulk of the electrons is readily dissipated in the lattice. On the other hand, when most of the positrons annihilate deep in the bulk of the lattice two relatively hard photons are emitted which can readily be detected several meters away from the specimen; this is a distinct advantage. They can traverse high pressure gases, the thick walls of a container and thus transmit information of an atomic or molecular nature, without leads or transducers, to detection equipment well outside the furnace, creep "stand" or vacuum enclosure.

However, one need not be limited to bombarding the specimen with the high energy positrons normally available. The new cadre of experiments involve thermalizing them in some solid and accelerating those which emerge through focussing lenses. They arrive at the specimen surface with a small tunable kinetic energy and are trapped at the surface even after slightly entering the solid. In this way, they become very strong non-destructive probes of surface conditions.

In this review, the author will try to put one interpretation on some of the conflicting data. To some extent, it represents a bias on his part. Whether it is correct or eventually supportable is another matter. The purpose is to focus attention on possible experiments which might lead to a better understanding. The second reason is to amplify the remarks by the speaker just before who had lectured on positron annihilation. He will undoubtedly have discussed the more traditional approach and traditional interpretation and to an extent, this reviewer has relied on him to cover much of the ground work and fundamental matters.

In this review, although twenty or more metals have been studied with positrons, two metals will be discussed in more detail than any others, namely iron and to a lesser extent, molybdenum. The technological significance of iron is self-evident. The interest in Mo is primarily because of its response to irradiation. The BCC metals have been considered at one time or another as candidates for the front-wall material for fusion reactors. While Mo may no longer be a candidate, a considerable amount of data on electron and neutron irradiated specimens has been accumulated and positrons have played a significant role in the interpretation of these data.

Also, in this review much more attention will be devoted to an analysis of positron lifetimes than to the Doppler Broadening and line shapes. Only passing reference will be made to the important subject of Angular Correlation since it is primarily a tool for measuring the momentum of electrons in a crystal and therefore for the determination of the Fermi surface. The latter subject does not figure prominently in the analysis of defects in metals and hence in the present topic of non-destructive evaluation.

Before launching into the meaty parts of the subject, another area of evaluating condensed matter non-destructively will be mentioned - namely problems of studying polymers and ceramics. At this stage, it seems to be easier and more productive to study the glass transition temperature in polymers than to study the densification of ceramic bodies. The pores left during sintering are too large in many cases to exploit the full power of this probe - that is, changes in lifetime with the number of atom vacancies in the cluster become saturated when the number approaches 125 - this could be a "spongy" region 15 nm on a side. The interesting size of a single condensed void in ceramics is about 1000 nm in diameter. The former size may not be easily seen with an electron microscope.

It is admittedly more traditional to discuss defects in metals, but the details of lattice vacancies and interstitials are well established and controllable in ionic materials. The electrostatic field associated with a specific defect is simple to calculate. Alkali halides being transparent have been studied by a wide variety of optical and resonance techniques. There seems to be a consensus even about the more complex color centers and structural relaxation effects. Possible traps for positrons can be evaluated even at positions which are not lattice sites (i.e. from the Ewald potential.) That is not so for metals, where there are several types of defects, such as vacancies, edge and screw dislocations, jogs or kinks. The macroscopic manifestations such as slip or cross slip, etc. are readily seen, but the atomic positrons near an edge or screw dislocation are not known with any certainty. This is because the force law between atoms is not well known. The Johnson interatomic potential (1) is used widely; it is a power series $\sum a_n r^n$ where the exponents n are positve numbers 1 to 6. A maximum range i.e. a cut-off at $r=r_c$ prevents the expression from diverging. The set of coefficients $\{a_n\}$ are adjusted to give a good fit to a set of experimental data. This power series is an oscillating function of interatomic distance. Various alternative interatomic potentials which have been used were recently discussed by de Hosson (2).

The atomic positions near the core of a dislocation are strongly dependent on the coefficients chosen for the power series. The relaxed positions of the atoms near jogs and kinks have scarcely been studied. But they may well be the ones that material scientists should be studying. This is the justification for starting with the ionic salts and primarily the alkali halides before going to the metals.

Positron Annihilation in Ionic Crystals

Positron annihilation experiments have been conducted with ionic crystals for almost 20 years. An Italian group, Bisi, Fiorentini and Zappa (3) observed that positrons passing through ionic salts decayed in several ways. The experiments of Brandt *et al* (4) to elucidate color centers in alkali halides, did lead to the observation that lattice defects influenced the mode of positron decay. The latest in a series of reviews by Dupasquier (5) was presented at the Varenna conference last year and the reader is referred to it for both the early and the most recent experimental results and interpretations as well as literature citations to other significant reviews.

It has been apparent for a number of years that the lifetime spectrum is complex and contains more components than the numbers of distinct lattice defects which are likely to be present at the same time in different alkali halides. This fact has stimulated much research but has not yet lead to a resolution of the problem. Bussolati *et al* (6) conducted a systematic study of 19 alkali halides from LiF to CsI. The specimens were single crystals of ordinary chemical purity i.e with low defect concentrations - occasionally large grained samples were used when single crystals were not readily available. For 14 of them, there were three components. The shortest lifetime, $@_1$, range from 137 for LiF to 313 picoseconds for KBr. This is for cases where at least 85 percent of the total intensity has been assigned - [the label here may be slightly different from the one they assigned], since they tended to call the shortest one $@_0$ when it no longer was the one which accounted for most of the decay. Even for the longest of the $@_1$ values the annihilation was by the low momentum electrons in the bulk of the solid. In a recent theoretical study by Kunz and Waber (7), the lifetime for annihilation in the bulk was calculated and found to be in excellent agreement with experimental data and these data are compared in Table I.

The second component $@_2$ ranged from 297 for LiF to 755 picoseconds for KBr. This assignment is slightly more tenuous because it is not clear whether it is properly $@_2$ or really $@_3$ for the bromides. If one restricts the list to the fluorides and chlorides, what is called $@_2$ represents 45 to 66 percent of the spectra, and the percentage of events assigned $@_1$ range

Table I Calculated and Experimental Lifetimes for Annihilation in the Bulk
(After Kunz and Waber)

	Calculated	Experimental
NaF	300 psec	297psec
NaCl	365	313
NiO	150	150
CaO	218	240
MgO	210	203

from 30 to 27. Study of the five iodides they worked with showed that the most frequent events had lifetimes from 327 to 425 psec with percentages ranging down from 85 to 53. For three of these five plus CsBr, the so called $@_0$ values are 126 to 183 psec. with percentages 26 to 15 percent. For LiI and NaI, the $@_0$ values are not listed and perhaps were not separated from the next component in the analysis of the data. The question is to what should this most probable lifetime component be assigned.

Somewhat more puzzling is the remaining 5 to 20 percent in the fluorides and chlorides. The values of $@_3$ range from 1210 (I=5%) for CsCl to 744 psec for RbF. For the high molecular weight set CsBr, KI, RbI and CsI the third longest component range from 686 to 980 psec. and with intensities ranging from 22 down to 4 percent. This subdivision has been made because they have the CsCl rather than the NaCl crystal structure although no reason is apparent why their lifetime spectra should have an additional component.

Then there remains the problem of the longest lifetime which is frequently called the tail with percentages amounting to 1 to 2 percent. The lifetimes range from 1870 (LiF) to 2430 (KCl) to 4220 (NaI) psec. Since these values are very hard to explain, experimenters have doubted that their computer fitting procedure is reliable enough to extract this low slope, low fractional intensity portion of the data when one makes a plot of the number of counts versus the lapsed time. The expression for the number of positrons remaining at time t is

$$N(t) = N(0) \sum I_i \exp(-t/@_i). \qquad (1)$$

Analysis of the "tails" have been distrusted and they have been ignored and/or not recorded by their authors. Three lifetimes are hard enough to explain why include the fourth and most inaccurate especially when it represents only a minor fraction of the annihilation events? That is a reasonable viewpoint.

The fourth lifetime is usually longer with a powder than with a single crystal of the same substance. The ratios $@_4$(powder)/$@_4$(crystal) are for example, (4050/2430) and (3600/1440). The pertinent intensity ratio, namely-I_4(powder)/I_4(crystal) is numerically smaller, i.e., 0.8/1.3 and 1.3/1.6. Times in excess of 1000 psec can reasonably be associated with the positron residing in a surface state - probably as positronium. This reflects the reviewer's opinion.

Let us turn aside from the speculation here and look back at the second component which for low molecular weight crystals is from 300 to 450 (and possibly up to 628 psec) For the heavier alkali halides the second component is 339 to 425 psec. Grouped in this way the range is small and the similar values suggest that $@_2$ might not be associated with bulk properties of the individual chemical species. This is of course not the conventional explanation; usually it is associated with cation vacancies. Although the vacant site will vary from a small size in the case of lithium to large for cesium; this observation would seem to require that the electronic charge density vary from only 1 to 1.5 arbitrary units to have the same lifetimes. It would seem possible, but if it is true, it is remarkable.

Before continuing this topic, it is worthwhile to review the lifetime spectra of various oxides and chalcogenides. For 17 compounds Bertolaccini et al (8) found that the short component $@_1$ ranged from 148 to 334 psec. For most of them the percentage was 85 to 10 percent. The second component, $@_2$ ranged from 315 to 910 psec. Most of the compounds were of the AB type but five were A_2B where A is monovalent. As indicated above these authors ignored the tail components $@_2$ or $@_4$. In their Table II, lifetimes and intensities for 27 alkali and alkaline earth halides are listed. These are more recent values than Bussolatti et al. The range of $@_1$ was 207 to 364 and for the alkaline earth halides 204 to 352. For the A_2B oxides, $@_2$ ranged from 315 to 584 for AB oxides 351 to 909 psec. For alkali halide $@_2$ ran from 297 to 685 and for the A_2B halides 347 to 538 psec. Without going into details, these values tally well with the earlier values if one ignores $@_0$, $@_3$ and $@_4$ components which would require a new labelling. That was done above. But it is not feasible here since the percentage I_1 and I_2 are not given - they might be inferred by matching the numbers preented by Bussolatti et al (6). The small range of $@_2$ discussed a few paragraphs earlier is neither confirmed or denied.

One reason Bertolacini _et al_ (8) constrained their data was because they used the two state trapping model and analysis. They assumed that at time t_o either:

a) all the positrons were in the state 1 and 2 characterized by different electronic densities; then I_1 would rep sent the state populations.

b) all the positrons are thermalized and migrating in the bulk of the crystal state and portions of them become trapped in State 2.

They present a method for getting the rate of transition from state 1 to 2 called k_{12}. Only case (b) permitted them to ascribe a common fate to the positrons annihilating in the wide range of some ionic and covalent salts they studied. For annihilation rates ($r_1 = @_1^{-1}$) is a linear function of the density of negative ions per unit volume, n*. This number is calculated as follows

$$n^* = v_- \ (pA/M)[1 - 4\pi R^3/3v_+] \qquad (2)$$

where p is the density, A is Avogadro's number, M is the molecular weight and the v's are the stoichiometric numbers in the formula $C_{v+}A_{v-}$. This model is based on a simplified picture of ionic solids namely that all the electrons associated with the cation C are found within a sphere of radius R_{cat} and that the electrostatic potential outside this radius of a cation is positive. Hence the positron would never penetrate R_{cat} and would only be annihilated by the electrons of the anion. The theoretical analysis discussed below yields plots of the probability of finding a positron along the shortest anion-cation line. Contrary to the hard ball model just discussed, the positron has appreciable probability of being found in a region where the electronic charge densities of both the anion and cation are also appreciable. That is the positron overlaps both regions with significant electronic density. That point is not greatly significant at this point in the review. What is interesting is that r_1 is well correlated with n*. They found that r_2 (or $1/@_2$) did not correlate with n*. It would seem quite plausible that it should if $@_2$ corresponds to the lifetime in a cation vacancy. Their argument assumes that the volume occupied by the cation sublattice are disjoint "regions inaccessible to the positron". A positron trapped in a hole (relatively speaking, in a negative region) would not have been scattered by the cations but would have better access to the electrons of the nearby anions.

The correlation they found was that $@_2$ is related to both the anion and cation populations and was relatively independent of the properties of the crystal namely

$$@_2 = (C_s/4)(pA/M)(v_+f_+ + v_-f_-) \qquad (3)$$

where f_+ and f_- are the effective numbers of electrons "available" from the cation and anion, C_s is Dirac fundamental interaction rate of a positron in the singlet state and the factor 1/4 corrects for the statistical weight of the singlet.

Independent of the details, this second decay mode is distinct from the first. They conclude that "the positron in the second state can be annihilated. Either with the positive ion or the negative ion, (a) the number of electrons per ion effective for annihilation seems to be higher when the ion size is larger" Also they conclude (8) that the positron wave function "in the second state must differ deeply from that in the first". They offer the observation that the transition rate (what we would suggest is the trapping rate) seem to increase with the formation energy of a Schottky defect. At this stage, any identification of $@_2$ with a positron trapped in a cation vacancy is inferential. But the point could be reinforced by looking at the behavior when positrons are implanted in crystals which have been doped to enhance the cation vacancy concentration. Interestingly, interstitial anions do not seem to have been considered in the treatments this reviewer has read. The case of AgBr would be desirable to study since Frenkel pairs are commonly formed defects. The effect of introducing dislocations by deforming the ionic crystals is another interesting topic since they are effective traps. Unfortunately, the results to date are not too convincing. Radiation by ionizing particles does so much damage that we will not use that information. In short, positron annihilation can be likened to a two-edged sword. It is sensitve to small changes in defect concentration but the experimental difficulties in characterizing the specimens tend to become the stumbling blocks to getting results reproducible by another research group and it does not lead to unequivocal interpretations.

(d) Thermal Effects in Ionic Crystals. The $@_1$ value in NaCl only increases by 13 percent when heated from room temperature to 973°K which is 100° below the melting point (9). When KCl was cooled to liquid nitrogen (77° K), $@_1$ increased slightly (about 10% due to cooling). Both $@_2$ and $@_3$ essentially did not change but I_3 was a few percent smaller at liquid nitrogen temperature then at room temperature. Measurements at the temperature of liquid nitrogen seem to show less regular changes than measurements made at room temperature (or 77 K) following a quench from high temperature. The resolved

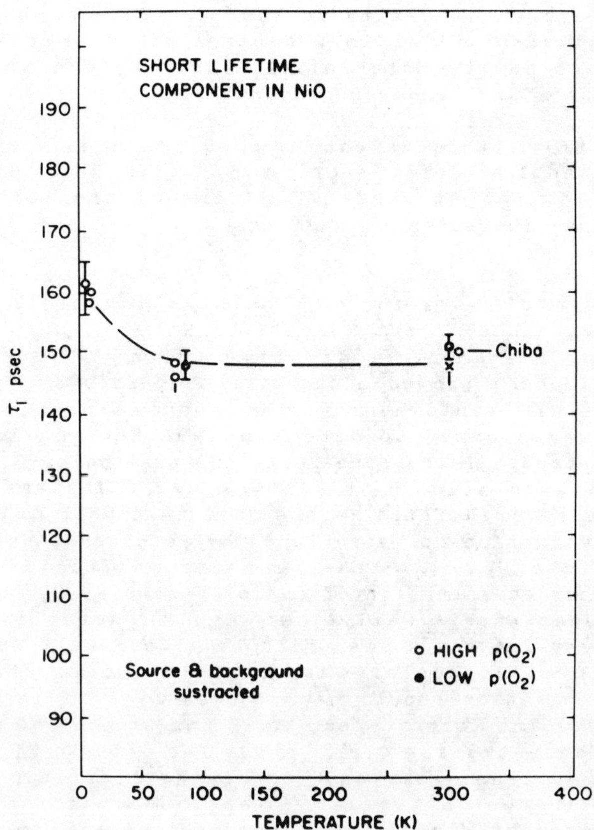

Figure 1- Variation of the shortest lifetime $@_1$ of monocrystalline NiO with temperature and cation vacancy concentration produced by annealing above 1200° C
(after Waber, Snead and Lynn)

$@_1$ (measured at room temperature) for KCl increased from about 185 psec when quenched from 550°C to 235 following quenching from 750°. For 77° K measurements the same values are 210 psec increasing to 260. The authors (10) kept $@_2$ = 464 and $@_3$ = 774 psec. This is scarcely strong evidence for thermal detrapping especially since I_2 decreased from 50 to 40 percent while $@_1$ was climbing.

It is not easy to increase the vacancy concentration above about 1 ppm by heating to 700° K. This number of defects is equivalent to minimum content of impurities in the typically high purity material. At 1000° C, thermally generated defects might reach 40 ppm.

Doping KCl with divalent Ca ions produces very little effect apparently - (See section 5.2.3 of Ref. 5). This statement does not strengthen the association of $@_2$ with trapping and annihilation at such traps.

This thermal behavior of the alkali halides is very consistent with the recent results of Waber, Lynn and Snead (10) on NiO. The shortest lifetime (mean value 150 psec) plotted in Figure 1 showed a negative temperature coefficient in the range of liquid-hydrogen to room temperature. The experiment was planned to determine whether shallow traps might be detected. This increase to 160 psec below 77°K might have that intepretation. The intensity I_1 was essentially constant. More signifcant is the fact that this high purity single crystal was one of a set used by Petersen et al (11) at Argonne National Lab to measure cation diffusion and the control of the concentration of cation vacancies by varying the oxygen pressure is well known (12,13). Actually, NiO is one of the very ionic systems which have been well studied by a variety of experimental methods. In short, the cation vacancy concentration should have increased from less than 1 part in 10^6 for 760 Torr oxygen, to 15 ppm at the low relative oxygen gas concentration of 1.6 ppm. The data in Figure 1 do not show significant changes from reannealing at T > 1200°C the crystal in the latter gas. The latter cation vacancy concentration should readily exceed any preexisting effects due to the small number of impurities in the sample. Cation vacancies seem to have very little effect on the lifetime. On the basis of the theoretical calculations the value of 150 psec for annihilation in the bulk NiO is in very good agreement with the experimental $@_1$ cited.

An intermediate component of approximately 350 psec has not been reported since the number is close to that of the polymer Kapton (14), namely 382 psec, which had been used to encapsulate the source. The relative intensity of this source was essentially constant at 22 to 23 percent throughout the set of experiments. It would be tempting to attribute this to annihilation in the para-positronium (or singlet state).

The component for a positron with a lifetime in excess of 1000 psec is plotted in Figure 2. Here some trends can be seen. There are positive temperature coefficients, the intensity increases by several percent. The error bars repre-

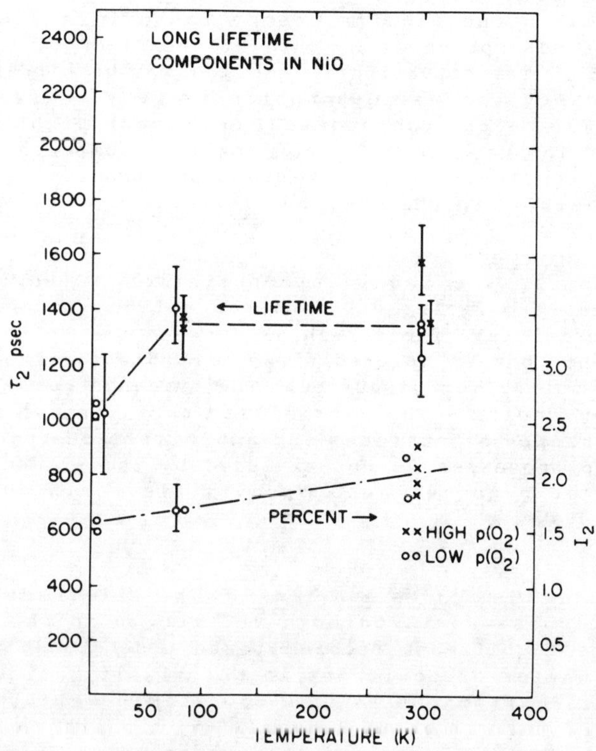

Figure 2 - Variation of both the lifetime θ_3 and the intensity I_3 of the same single crystal with temperature and oxygen pressure. (After Waber, Lynn and Snead)

sent variations between several determinations on the same specimen in the same annealed condition. Then data were not corrected for any source - surface contribution. This investigation of NiO does not give any evidence of thermally activated desorption or shallow traps. There are two possibilities, one that almost all of the positrons are localized in deep traps or that diffusion was not sufficient for most of the positrons to reach the traps.

A recent calculation of the binding of a positron to a cation vacancy was made by Kunz and Waber (16). The binding energy was 8.1 eV in NiO and 9.1 in CaO, but only a fraction of an eV in several alkali halides such as NaCl. It was felt by the authors that because the latter cluster calculation was extremely sensitive to the basis set of atomic orbitals used, the results were unreliable. This nickel oxide value was consistent with the binding energy found in a theoretical study of the adsorption of H onto NiO carried out by Surratt and Kunz (16). The situation is ambiguous, the binding energy is so high that thermal desorption is very unlikely to be observed. The second (unreported) component might be due to annihilation in these traps. The long-life component is probably due to the annihilation of quasi-positronium (qPs) trapped in the bulk or in the surface of NiO.

In short, the usual interpretation of annihilation lifetimes and broadening dependent on cation vacancies trapping at internal interfaces such as between the colloidal metal and the host in colored and/or irradiated alkali halide is possible but seldom discussed. The evidence seems to be far from convincing. The possibility that many of the positrons are trapped at surfaces and annihilate due to pick-off or spin-flip processes is not excluded by the evidence. That must still be regarded as a speculative position for the reviewer to take.

(b) Properties of Ps and qPs. The lifetimes of Positronium and quasi-Positronium have emerged in the last few years as the significant problem to be understood in discussing the behavior of positrons in the alkali halides and in the oxides. Positronium Ps is a low atomic weight hydrogen atom. An electron is bound "loosely" to a positron (or anti-electron) which replaces the proton in normal hydrogen or the deuteron in heavy hydrogen. A proton weighs roughly 1850 times as much as an electron, whereas a positron weighs no more than twice as much. Much of our intuition and insight into materials problems can be brought to bear on the treatment if we think of a positron as a lightweight hydrogen atom. This would be convenient if the "darned thing did not die so soon". The death of positronium is violent just as is the death of one of the partners, the positron. The two or three gamma rays which are created are the "fingerprint" of what conditions were like when the Ps atom self-annihilated. If the spins of the electron and the positron are anti-parallel as they are in the para- or singlet state, two photons are created. If the spins are parallel then an odd number of photons are required to conserve the momentum of the ortho- (or triplet state) positronium. The spin situation is strictly parallel to the case of ortho- and para-hydrogen. Direct transition from triplet to singlet state is forbidden quantum-mechanically. Several things can mediate the transition. Another electron can impinge on the Ps and pick off the positron to form the lower energy singlet state and allow the former electron to drift off. A spin-flip may occur when the Ps atom collides with a free surface or with the wall of the void, particularly if a paramagnetic ion is present near the point of impact. Third, a moderate magnetic field will influence the transition and quench the long lifetime of the orth-Ps. We will not go into either the kinetics of Ps or qPs formation or the comparative lifetimes at this point.

If this species has a lifetime which is long with respect to the characteristic time for an atomic vibrations (which are of the order of a 0.1 to 1.0 psec) then we can think of it as light atomic hydrogen and we can make some predictions which can, of course, eventually be tested for correctness.

There are a large number of metals which chemisorb hydrogen - why not Ps? There are a large number of metals in which hydrogen will dissolve and migrate - why not Ps? Here one distinction is important to make as Dupasquier (17) does; - the term positronium or Ps can be reserved for a free e^+-e^- bound pair and when the bound pair interacts with a solid medium, he used the phrase "quasi-positronium" which was introduced by Held and Kahana (18). The previous question should be modified - why isn't it reasonable to expect qPs to reside comfortably in those metals which tolerate the dissolution of atomic hydrogen? The charges are the same in H and in qPs; one difference is the mass of the particle. Thus one would not expect the binding of qPs to be as strong so qPs could diffuse more readily. There is little serious discussion that the hydrogen atom exists as a bare proton when dissolved in a metal. Instead it is generally thought of as a screened proton, i.e. one screened by a cloud of electrons. The detailed quantum mechanical, band calculations of metallic hydrides carried out by Switendick (19) are incompatible with H having a positive charge in such compounds and many hydrides look as though H^- is a more realistic species. The details depend on precisly how one divides up the total space into the volume for the hydrogen atom/ion and the volume for the metal.

There is a further piece of information namely that in electromigration studies of iron, i.e. diffusion assisted by a massive current of low voltage electrons - virtually an "electron wind" - the drift is compatible with the hydrogen having a charge of +0.25 electron units (20,21). Returning to the state of a positron, one would reasonably expect the e^+ to exist as qPs if here we think of an electron cloud around the positron and that a small apparent charge of either + or - may be associated with qPs. In fact, Mills (22) reviewed the experimental evidence of qPs^-. To summarize the analogy, one predicts that qPs will be found adsorbed on many solids as well as being absorbed into them. Further, that various chemical compounds would be anticipated. If oxygen is present in the solid as an oxide, the compound can have hydrogen present as the OH^- anion. By analogy, $O\text{-}Ps^-$ might be found perhaps infrequently.

Unless very strong positron beams are developed it is unlikely that one would find two Ps atoms, even two positrons, simultaneously in that small volume.

The early thinking about qPs was that near a cation vacancy, the positron would bind to one of the neighboring anions to form a covalent bond such as in the OH bond. The e^+ would not have to bring its electron along with it. In this covalent picture, the qPs species would be highly localized. The first question was - wouldn't it be readily shared by the six or eight anions which surround the cation vacancy, and thus be more delocalized than one would have in a simple O-H covalent bond? But if one pushes the atomic hydrogen analogy slightly farther, it is not hard to see that qPs might well become "delocalized" and migrate past ions or even atoms at quite a number of lattice sites. Hydrogen atoms diffuse readily in ionic and particularly in metallic lattices. In fact, the very important technological problem of hydrogen embrittlement;. which is well known to occur in steel, stainless steel, and several BCC metals, is discussed, for example, in the recent review by Thompson and Bernstein (23). It would not occur if atomic hydrogen did not migrate readily many hundreds of lattice parameter distances before it became trapped at voids, dislocations, interfaces, etc. (probably as H_2 or CH_4). There are never enough Ps atoms and they annihilate too rapidly for "molecular" qPs_2 to be detectable. This does not rule out the idea that qPs may migrate over considerable distances as a neutral atomic species before being trapped and annihilated. In some ways, that might be a more realistic statement of condition for thermalized positrons - in contrast to the high energy positrons which impinge on a solid from ^{22}Na or ^{58}Cu.

Here it seems important to be a little more specific about what we mean by the term "delocalized Ps" which Dupasquier (17) uses. He is careful to distinguish two cases

a) The wave function $\psi(\mathbf{r}_p,\mathbf{r}_e)$ of qPs is significant only over a region which contains zero or at most few atoms of the solid.

b) The wave function $\psi(\mathbf{r}_p,\mathbf{r}_e)$ [or using his notation $(\mathbf{r}_+,\mathbf{r}_-)$] does not die off rapidly with distance and is signifi-

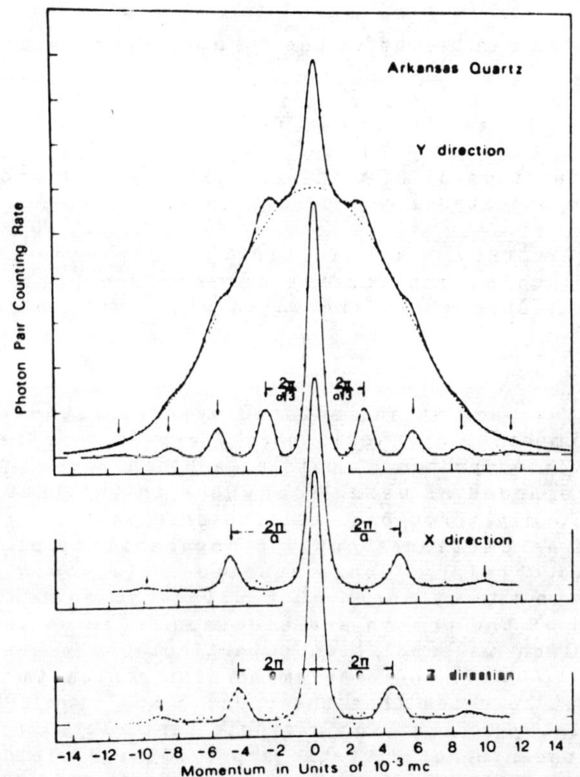

Figure 3 - Angular correlation of gamma rays fom two-photon annihilation of positrons for the three principal axes. Smooth dashed curve has been subtracted before data for X, Y and Z directions were plotted. (after Hodges, McKee, Triftshauser & Stewart)

cant over a region with dimension L containing a large number of unit cells of the solid.

He does not initially assume that this delocalized qPs wave function must be a Bloch wave. But he makes the statement in his section 1.8.2 that "the interaction of a correlated e^+e^- pair with a periodic lattice is invarient under translation of a lattice vector, the total crystal momentum hK of the pair is a good quantum number to label the eigen-states of the system and the corresonding energy eigen-values E(k); [by] changing coordinates from (r_+, r_-) to

$$\mathbf{R} = (\mathbf{r}_+ + \mathbf{r}_-)/2$$
$$\text{and} \quad \rho = (\mathbf{r}_- - \mathbf{r}_+)$$

the wave function can be put in the form of a generalized Bloch function:

$$\psi(\mathbf{r}_+, \mathbf{r}_-) = \exp(i\mathbf{k}.R)\, V_K(\mathbf{R}, \rho) \tag{4}$$

where V_K is a function of **R**, of ρ and possibly of **k** and having the lattice periodicity of the direct lattice or R space."

Although $V_K(\mathbf{R}, \rho)$ was initially restricted to a few unit cells, by this equation the interactive potential is assumed to exist throughout the solid with lattice periodicity.

Looking back at an isolated hydrogen atom in a solid, although it is mobile, one would hardly expect to effectively treat the correlated proton-e^- pair as a Bloch wave. The pair is generally regarded as being somewhere in the lattice and not spread uniformly throughout all the unit cells. He emphasizes that "no assumption about the separability of R and . is necessary to obtain ..." this equation. It would be just as easy to set up the hydrogen atom using the mean distance and separation of the proton and electron. *Per se* that does not make the Bloch wave solution a particulatly tractable or attractive way to treat the problem of positronium in a solid. However it might be noted that the bound state atomic problem reduces essentially to that for a single particle with reduced mass and that the non-relativistic approximation yields solutions with essentially the hydrogen atom wave functions.

This idea of the positronium existing as a Bloch wave occurs in several recent papers. The greatest justification comes from the angular correlation curves, the discovery of Brandt, Coussot and Paulin (24) of "side bands in addition to narrow central peak on the two dimensional plot of the angular correlation curve from a quartz crystal these satellite peaks occur at reciprocal lattice vectors." Their results were confirmed by Greenberger _et al_ (25) who also investigated the influence of a magnetic field in quenching certain components Hyodo and Takakusa (26) have reported evidence for qPs occurring as Bloch wave in alkali halides. Additional discussion was recently given by Manuel (27). Impressive as this evidence is, it is not conclusive proof of the Bloch wave nature of qPs. A typical plot of the annihilation counts versus elapsed time is presented in Figure 4.

The interpretation given above ignores the contribution due to the valence electrons of the host or solvent lattice; Hodges, McKee, Triftshauser and Stewart (28) point out that it is they which have the periodicity of the lattice

> "the above approach has a serious defect in that it does not attempt to include the crystal valence electrons in the system and in particular neglects the Pauli indistinguishability between the valence electrons and the electron of the positronium".

The latter authors were successful in describing most of the experimental results by calculating the pick-off rate due to the valence electron orbitals (r_e) interacting with the qPs orbital (r_e, r_p) in addition to the self annihilation of qPs.

Dupasquier (17) devotes a lot of discussion to the density parameter **K** where kappa is dimensionless and defined by the equation

$$|\psi(0)|^2 = \kappa \, |\psi(0)|^2, \qquad (5)$$

The left side is for qPs whereas the quantity on the right hand side is for free Ps and represents the average electron density experienced by the positron (taken as a point charge). The values of **K**, which Hodges _et al_ deduced from the data of Greenberger _et al_ was 1/2; Berko is quoted by Duspasquier as finding **K** < 1. Apparently Gambarini and Zappa (29) believe that **K** = 1. It is quite reasonable that qPs is "swollen" compared to free Ps. The effect of the dielectric constant of the host lattice will reduce the electrostatic potential between e^+ and e^- and thus reduce **K**.

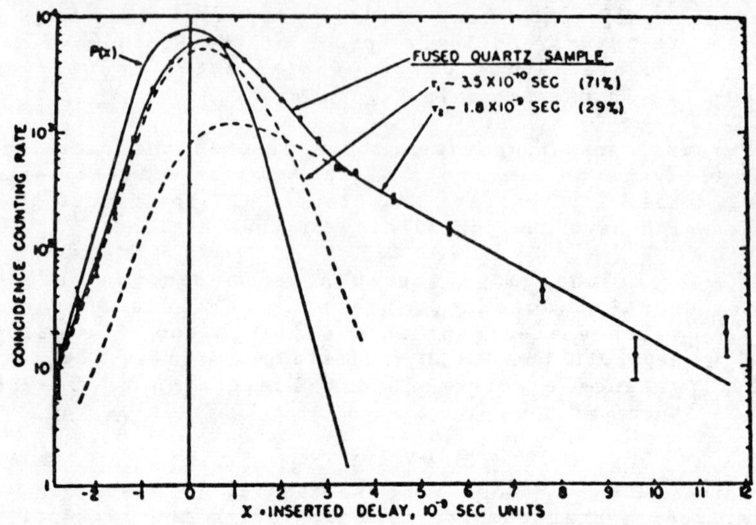

Figure 4 - Counting rate for three photon annihilation plotted versus time delay(in nanoseconds).Analysis based ontwolifetimes in Equation 1.

It is interesting that qPs is what might be called "Lite Hydrogen" in verbiage used currently used in advertising campaigns. The screened proton in the hydrogen atom and the screened positron in qPs will be trapped in the same way and in much of the same traps. The binding energy will be smaller for the qPs since the binding energy scales as the reduced mass, i.e. will be half as large. This means that the qPs will probe the deeper and the more important of the H traps.

The satellites occurring at reciprocal lattice vectors of quartz are evidence of the pick off reaction between the valence electrons and the relatively localized qPs occurring in the bulk of the lattice. The same thing will occur if the qPs is in a bound (i.e. chemisorbed) state on the surface of the solid and interacts with valence electrons which can be described by two dimensional Bloch waves.

The lifetime which Hodges et al [28] estimate for the pick-off annihilation is 320 psec for quartz. Gambarini and Zappa[29] got 379±15 psec with the high intensity value of 42 percent in alkali halides. Thus some of the values of θ_2 of comparable magnitude would seem to be compatible with the annihilation of the ortho-Ps in solid resulting from the interaction with valence electrons. Before closing the section a word of caution is in order. The wave function referred to above which describes the valence electrons as well as the Ps atoms "... ignores dynamic correlation effects between the positron and valence electrons, which would given enhancement of the pick-off rates..."(28). Treatment of the response of the electron system to motion of the positron and its electron is quite difficult and beyond the scope of this review. For studying the surface and bulk defects in a metal, the elementary interaction above of valence electrons, as well as the conduction electrons, with the qPs atom and/or with e^+ is sufficiently refined. We will not go further in this discussion of qPs.

Characterization of Defects in Metals

Positron annihilation is one of the succesful methods for studying the various defects in metals. First it is non-destructive. It is sensitive to defect concentrations of 1 part in 10^7. They are trapped there and have greater lifetimes and reduced widths (FWHT) of their Doppler Broadening Curves. It is sensitive to void sizes which cannot be easily resolved by transmission electron microscopy (TEM) or small angles neutron spectroscopy (SANS).

Positron annihilation analysis has not yet arrived at the full potential it has largely because the specimens have not been well characterized and even when the composition or preliminary heat treatment is known, the specimens are subjected by workers to signicant plastic deformation such as 10 to 40 percent reduction in area. This produces vast numbers of dislocations which interact with each other to form planar arrays or more likely, dislocation forests with jogs and other defects such as vacancies, subgrain boundaries, etc. It has taken quite a few years to untangle the interactions of interstitial atoms such as carbon and nitrogen or oxygen with

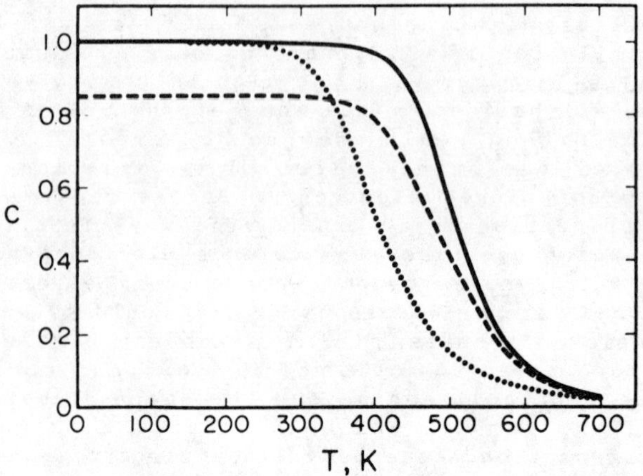

Figure 5 - Variation of the concentration of hydrogen remaining in trap sites with an enthalpy of 58.6 kJ/mole. c_o established by - quenching in a closed system from 700°K; trap density = 9 x $10^{15}/cm^3$ shown as soldi line. The (---) line corresponds to 100 x 10^{15}/cc. The (...) line corresponds to the latter c_o but quenched in an open system. (after Hirth)

vacancies and interstitial but interactions with dislocations in iron was easier to investigate. Progress on the formation of hydrogen-interstitial pairs in BCC metals has come mainly from work on other BCC metals using internal friction methods. It will take this same type of careful preparation of specimen and control of experimental conditions before the interaction of positrons with the contained or with intentionally inserted defects and/or intestitial-vacancy pairs can be unravelled. The clarification can be greatly facilitated by the positron and qPs annihilation, because these species get trapped at the defects and sample the conditions locally and send back information from just where it is needed.

Hirth (30) in his recent review on hydrogen embrittlement lists ca 20 kJ/mole as the binding energy of hydrogen to screw dislocations and 58.6 kJ/mole for mixed edge dislocations. These are deep traps. Apparently a value was not listed for binding to a vacancy. The thermal detrapping of hydrogen from traps with the binding energy of an edge dislocation is shown in Figure 5 for two different trap concentrations.

(a) Interaction with Dislocations- Granatelli and Lynn (30) reviewed positron annihilation in metals recently. A number of studies have been carried out with purified metals after it was observed that a second lifetime longer than the bulk value $@_1$ occurred with plastically deformed samples. These subsequent studies have indeed confirmed this observation. Positrons are trapped in metals which contain a large number of dislcoations and the lifetime is 20 to 40 percent larger. It is generally observed that the lifetimes associated with trapping by vacancies and by dislcoations have very similar valus. For example, the annihilation in the bulk gives 110 psec for iron and 165 for dislocations and 170 psec for vacancies (32). None of these values is more precise than 5 psec due to the difficulties associated with fitting the data with two and possibly three exponential terms in Eq. i. So the values given for the vacancy and the dislocation could be the same. The point being made here is that it is difficult to prepare plastically deformed metal specimens which contain primarily one type of defect, and in general speaking, efforts were not made to investiate with transmission electron microscope (TEM) which kinds of dislocations were present and how many there were.

It is a general observation that the annihilation rates for vacancies and dislocations are very similar in many metals. This had lead a number of workers to suggest that the trapping is indeed weak in a dislocation, and that the observed annihilation event may actually be associated with a jog or a kink in the dislocation. Granatelli and Lynn state that "Since screw dislocations are not generally believed capable of trapping positron the ideal situation [for unscrambling the effects of different types of dislocation interactions] would be the production of a high concentration of long straight edge dislocations in an otherwise perfect single crystal. In this way, the annihilation parameters associated with a dislocation could be established" (31). It is reasonable that the positron will be found in the dilational side of an edge dislocation in much the same way that a hydrogen or carbon aton is trapped on this tension side, i. e., below the slip plane where the half plane of atoms is missing

Jogs occur when two dislocations "intersect" each

other while moving on different active slip planes. This is a region of considerable relaxation or atomic rearrangement - it could have a vacancy-like character.. Nucleation of double kinks is generally regarded as a necessary step for lateral motion of a screw dislocation, Hirth (30) indicated that the hydrogen trapped by screw dislocations can be 'found" in such kinks. It would be quite reasonable to expect that the positrons would be found there as well. At first, this looks like a stumbling block, but it may be the most useful feature of positrons, if they can indeed give a measure of the number of kink or jogs present in a sample since these are difficult quantities to measure and corroborative information from a different source is sought and highly regarded. So far experiments have not been well organized to study this type of possible trap.

(b) Thermal Behavior and Trapping Models - Smedskjaer, Manninen and Fluss (33) have proposed a model which incorporates two related defects with different binding energies. The straight edge dislocation would barely trap the positron and delocalize it, whereas any jog along the dislocation would bind it strongly enough that very little thermal detrapping would be expected. They have explored the temperature behavior of the model and find that they can explain many effecs which are observed. The earlier proposal of Hjelmroth *et al* (34) involving type A and type B traps has many general features in common with the model just mentioned.

The traditional way to study the variety of defects produced by electron or neutron irradiation is to anneal the specimen isochronally at progressively higher temperatures. Several stages are seen. Stage III will be discussed below. During recovery stage of recrystallization, at least some of the dislocations become free to move - some are destroyed when for example, dislocations of opposite sign meet. Others migrate to the grain boundaries, or form polygonal walls. In general, there can be a reduction in the number of dislocations "seen" by the positron as it becomes thermalized. Consequently, there is rduction in the observed lifetime, since most of the annihilation events will occur in the bulk of the crystal.

The problem of relying on isochronal anneals is that a number of phenomena are occurring simultaneously. For example, the data of Hautojarvi*et al* (35) in Figure 6 illustrate how complex it may be to analyze and interpret the results of isochronal anneals. These results will be discussed below. One of the complicating factors is that, recrystallization has been observed to occur during the recovery process (36,37). There is evidence that the behavior during annealing is related to the size (38) of the stacking fault energy, E_{sf} . The long range stress fields of cell walls will induce cross slip of dislocations (cell structures form more readily with high stacking fault metals) will facilitate dislocation climb during annealing with attendant annihilation of the dislocations and rearrangement. Cotterill and Mould (35) point out that when E_{sf} is high as it is with aluminum, recrystallization will precede recovery except at low strains. In contrast, "..little mechanical recovery is observed for copper or its alloys prior to recrystallization." This is not the place to go into the various features of the energy release processes which occur during recrystallization. It appears to this reviewer that an improvement in understanding could come from careful experiments carrying out isothermal anneals using positron annihilation as a non-destructive investigative tool.

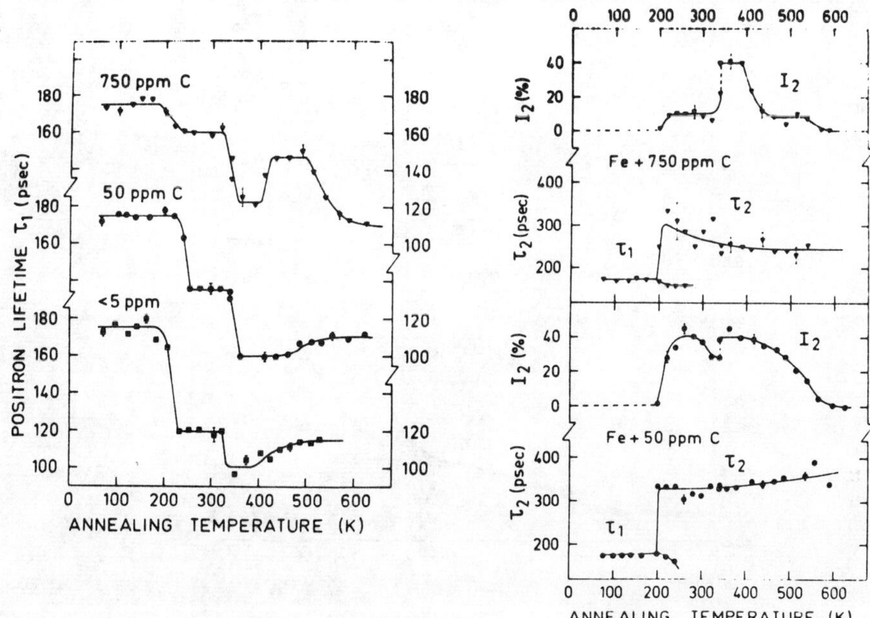

Figure 6. Variation with Annealing Temperature (a) of the shortest lifetime $@_1$ for three carbon contents and (b)of $@_2$ and I_2 for carbon additions to the high purity iron. (after - Hautojarvi et al)

Vacancies are reported to become mobile in pure iron at temperatures as low as 210° K (30). This interpretation would appear to be unlikely. Schaefer et al (39) argue from the temperature behavior and the fact that the enthalpy of migration exceeds 1 eV. that monovacancy migration occurs well above room temperature in the range of 450 to 600° K. They suggest that the increase in positron trapping and the recovery, which is "..in many respects the analogue to Stage III in fcc metals " ...must be due to long-range migration of an interstitial type defect," since "it cannot be due to the formation of vacancy clusters.".

The work of Sato and Meshii (40) on ultra-high purity iron and on iron-carbon alloys suggests that self-interstitials are released from trapping carbon atoms in Stage III or > 250°K. The migration energy is 0.53 eV which is higher than the accepted value for the migration of self-interstitials (41), namely 0.33 eV. They believe that there is no stage which corresponds to vacancy migration below 450°K.

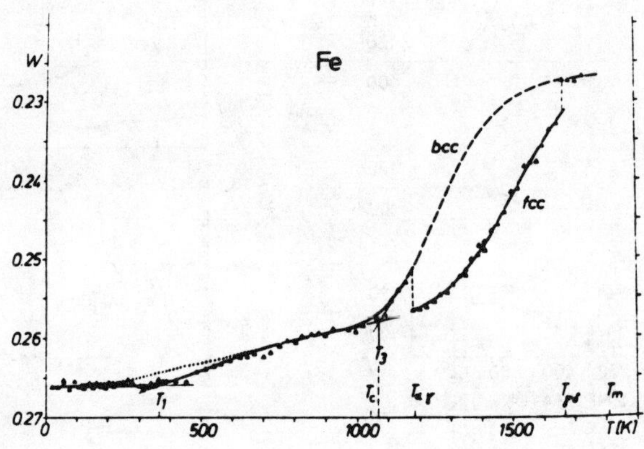

Figure 7. Variation of the Line Shape Parameter with Meaurement Temperature for High Purity Iron. The several Phase Transformations are marked with Arrows. (after Schaefer et al)

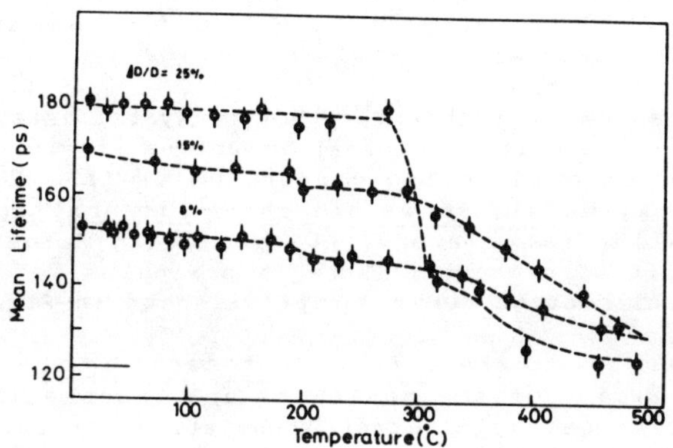

Figure 8. Variation in the lifetimes observed by Doyama and coworkers after 5, 15 and 25 percent deformation of copper.

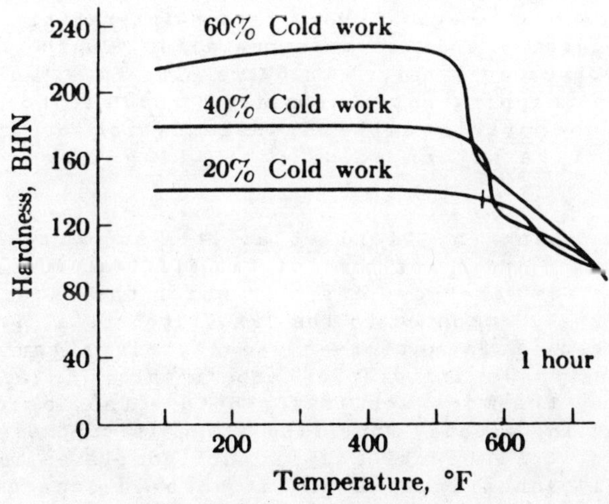

Figure 9. Softening of 65Cu35Zn by Recrystallization (after ASM data)

One important aspect of positron annihilation should not be overlooked in connnection with such annealing studies and that is that the positron is not directly sensitive to self-interstitials (or increased density regions) but they detect low density regions such as vacancies, voids or dislocations. Said slightly differently, they sample only one part of a Frenkel pair.

It is commonly observed that recrystallization does not occur unless there is a critical amount of strain. The data of Doyama (42) on cold rolled copper, which is a low stacking fault energy metal, shows the change in positron lifetime with annealing temperature. At low strain values, little - change in lifetime occurs. There is a similarity to the typical recrystallization curve for brass shown in Figure 9.

Returning to the work of Hautojarvi et al in Figure 6 a large percentage of the positrons are trapped in the monovacancies produced by electron irradiation. Annealing through 220°K splits the lifetime into two components. The long lifetime of ca 300 psec is "..strong evidence for vacancy migration which results in agglomeration of small three-dimensional microvoids." These authors concentrate on interstitial-vacancy interactions. Even in nominally pure iron they conclude that there is a significant number of carbon-vacancy pairs since positron "..trapping only into [bare] microvoids would give $@_1$ = 75 psec at 280°K." It is the "..migrating vacancy which is captured by a stable carbon atom." The activation energy they quote for carbon migration is 0.86 eV. The second annealing stage near 350°K, where the $@_1$ values drop considerably - this is the temperature region where carbon-interstitial pairs are known to migrate, This is interpreted as "further decoration of the carbon-vacancy pairs" which results in a "nullification of positron trapping into the pairs." Such a short lifetime $@_1$ is not compatible with any calculation or independent experimental result on voids with which this reviewer is familiar.

The paper by Diehl et al (43) supports one of the views above and indicates some of the difficulties encountered - namely, "that in those cases in which the vacancy concentration is small compared to the IFA [intersitial foreign atom] content direct information on vacancy migration cannot be obtained.." Even in irradiation experiments, "..vacancies are decorated by the more mobile IFA before they become mobile themselves." The vacancy annealing stage is suppressed and the dissociation of interstitial-vacancy complexes compete with the "normal" annealing out of undecorated vacancies in the same temperture range. An alternative interpretation which apparently was not considered is that the carbon atoms are interacting with mobile screw dislocations and their kinks or jogs, and these look very much like vacancies.

Seegers et al (44) showed that for high purity iron the positron-dislocation binding energy is 0.63 ± 0.03 eV. In a further study (45) which involved isothermal annealing, the activation energy for migration of interstitials such as carbon and nitrogen was found to be 0.53 ± 0.07 eV`.

Coworkers with Petersen (46) have proposed a four-state trapping model. The relative rates are shown in Figure 10, the lowest I_{13} being for voids. This model permits a transition at a reasonable rate from either the bulk or at a rate k_{22*} from a shallow trap such as an edge dislocation or vacancy loop into State 2* associated with a jog. Here the binding energy is strong enough that the detrapping rate k_{23} is negligibly small. Each of these states has a progressively longer lifetime.

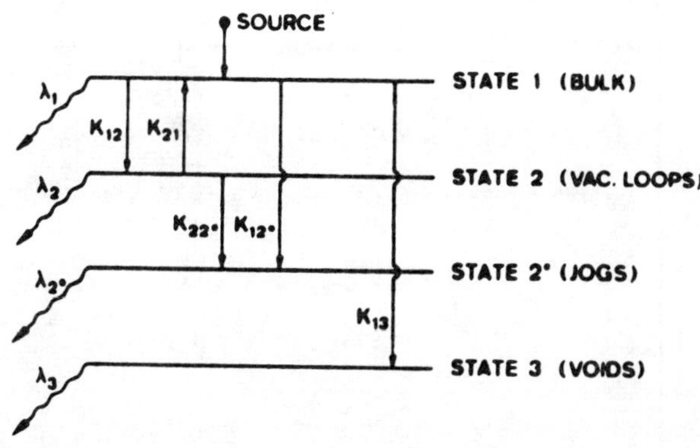

Figure 10. Four State Trapping Model assumed appropriate for Neutron Irradiated Molybdenum. (after A. Pagh et al (46))

Following irradiation, vacancies and interstitials coexist. When annihilation occurs, it is not possible to distinguish whether it was the vacancy or the interstitial which moved into the vicinity of the other defect. However, Nielsen and Petersen (47) had an interesting suggestion, which is based on the well documented effect of the number of sites in a vacancy-cluster on the lifetime of the positron. As indicated in Figure 11, the size of a microvoid may increase or decrease depending which species is mobile at the temperture. When the monovacancies are moving, the voids will grow and both I_2 and θ_2 will increase. It is usually explained that the intensity of the second component I_2 increases with the number of such defects per unit volume. They produced many Frenkel pairs by irradiating molybdenum with 10^{15} electrons/cm^2. During stage III recovery at 480 to 560°K, they deduced that there was an increase in void size and a reduction in $@_1$ which is associated with vacancy-loops. Both results are compatible with vacancy migration during this annealing stage. After further annealing, the positron annihilation data returned to values the specimens had before irradiation.

It is usual to connect the value of the intensity I_2

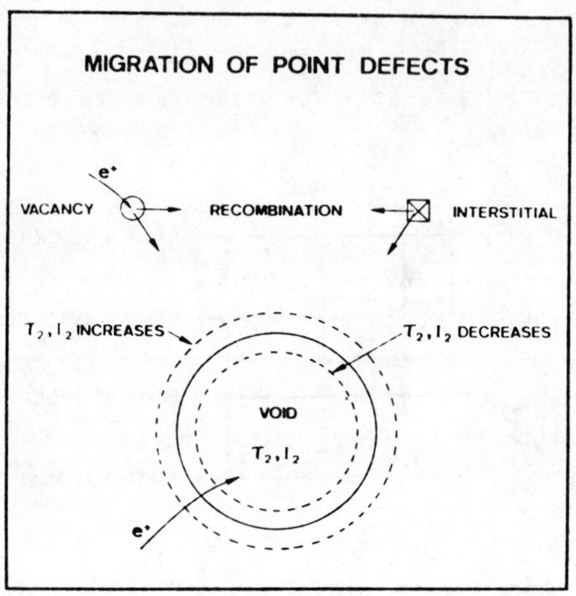

Figure 11. Schematic Representation of the Migration of Point Defects in a Metal and their Effect of the Size of the Microvoid. (after Nielsen and Petersen)

with the number density of microvoids. An alternative suggestion is that I_2 might correlate with the surface area of the voids present. This would be reasonable to expect if the adsorption of the positron onto a surface of the void is a precursor of the annihilation event.

Hydrogen was mentioned above because hydrogen embrittlement is a very important technological problem and from the large number of papers published each year it is eminently clear that hydrogen interacts with various defects in iron or other metals in a complicated manner. The interaction with positrons will be similar.

Nielsen et al (48) recently performed an interesting experiment on the effect of hydrogen in steel. Irradiated samples where annealed to produce vacancy loops and hydrogen was implanted by an accelerator or injected cathodically. One lifetime, approximately equal to 165 psec was the response from the loops and dislocations. The other θ_2 in the range of 300-400 psec was identified with voids. With hydrogen added, there was a reduction in the intensity I_2 suggesting that hydrogen trapped in the voids was interfering with the trap-

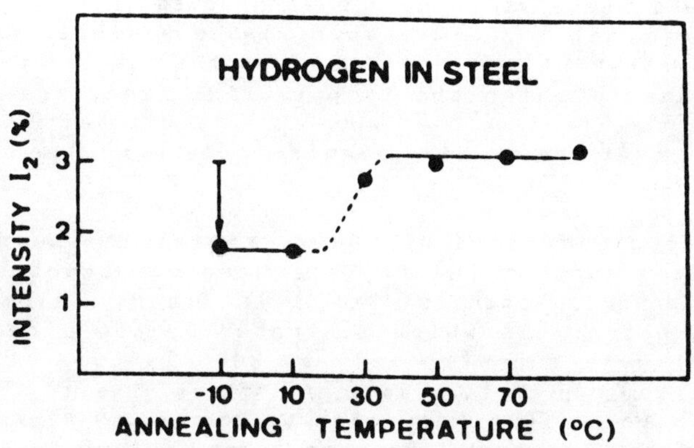

Figure 12. Effect of Isochronal Annealing after Hydrogen Charging of Irradiated Steel Specimens on the Intensity of Longer Component. Detrapping increases the Annihilation in the Voids. (after Nielsen et al)

ping and annihilation of positrons in the voids. Changes in the intensity of this component is shown in Figure 12 as the hydrogen is thermally detrapped during isochronal annealing. There is some similarity to Figure 5 if one plots (1 - c).

Theoretical Aspects of Trapping in Defects

The effectiveness of using positrons as non-destructive probes is linked intimately with their dwelling some time at various defects which have reduced electronic density. What theoretical evidence is there that they would become localized?

To assist the reader in visualizing the potential which causes a positron to become localized and how certain kinds of defects influence this potential, a digression is desirable in which the specific steps involved in creating such a crystal potential are indicated. Following this digression, we will return to the main topic of how such defects influence the positron lifetime.

In Fig. 13 a perspective plan shows what the numbering system is and where atoms have been removed from a face centered cubic metal to form a step on the surface. The basic dimension is that 10 units are equal to the lattice parameter, a_o. The plane in which the isopotential lines are 1.5 cell edges above the free surface; when it is indicated in the next few figures that the viewing planes is the 15th, the height is not $15a_o$.

The contour plots of the potential in the next two figures for a step on the surface were constructed by the method of Kennard, Waber and Tsui (49). Briefly, atomic self-consistent potentials which generated by the Dirac-Slater method (50) were superimposed at 8000 lattice sites. The individual atomic potential at some lattice site $\mathbf{r_i}$ slowly approaches zero at 60.0 Bohr units (a Bohr = 0.0592 nm) as is appropriate for a neutral free atom. Many Coulomb tails from the free atoms which are placed at the lattice sites contribute to the potential at any point $\mathbf{a_j}$ in any cell. i. e., at any point in the 10x10 grid drawn in the unit cell, Values of the electrostatic potential were interpolated from those which were calculated on a "logarithmic" radial mesh. In the same way, the self-consistent electronic densities from the 8000 atoms which surround a grid point were interpolated from the set of wave functions and superimposed at $\mathbf{a_j}$. These two sums

$$V_{el.cryst}(\mathbf{a_j}) = \sum V_{atomic}(\mathbf{r_i} + \mathbf{a_j}) \qquad (6)$$

are the electostatic crystal potential and the crystal elec-

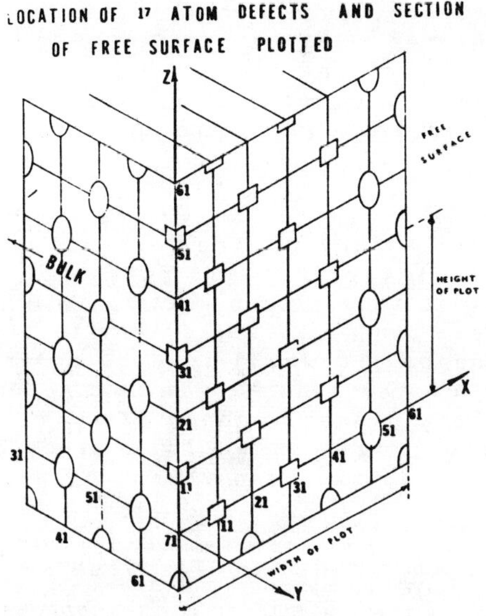

Figure 13. Explanation of the grid and numbering system for a FCC metal. Removal of the 17 atoms indicated on the (010) free surface will form a step.

tronic density, The latter is usually written as

$$P_{cryst}(\mathbf{a_j}) = \sum P_{atomic}(\mathbf{r_i} + \mathbf{a_j}) \tag{7}$$

and the sums were tabulated for the set of several hundred grid points a_j, that is at points well above and well below the surface. To do solid state calculations, it is necessary to include the exchange potential. The local exchange was obtained by taking the cube root of the summed charge density at $\mathbf{a_j}$ and multiplying it by a suitable scaling or Xa constant, such as two-thirds. Thus, the crystal potential can be written as

$$V_{cryst}(\mathbf{a_j}) = V_{el,cryst}(\mathbf{a_j}) + 2/3\sqrt[3]{P_{cryst}(\mathbf{a_j})} \tag{8}$$

It is a costly procedure to create this table of crystal potentials for several thousand grid points, namely those which run over several cell dimensions parallel to the surface in both directions and in addition span the space above and below the free crystal surface. However, once it is done for a perfect crystal of a given element, it does not need to be redone, when one wishes to treat a defected crystal. Instead, one has only to subtract the atomic potential and the electronic density of each lattice site $\mathbf{r}_{\mathbf{m}}$ which is to be removed and then add back in the equivalent quantities for the new positions $\mathbf{r}_{\mathbf{n}}$.

The potential variation above the surface when a step occurs, is shown in Figure 14. The fourfold symmettry of the underlying lattice is evident and no sharp discontinuities occur at the edges of this planar defect. This potential would repel electrons even though it would lead to the trapping of a positron, consequently it would lead to an increase in the lifetime, because of the reduced electronic density This defect is similar to a planar vacancy cluster. Note that the potential near the nucleus would be very large on this scale, i. e., it would take many sheets of paper to display it.

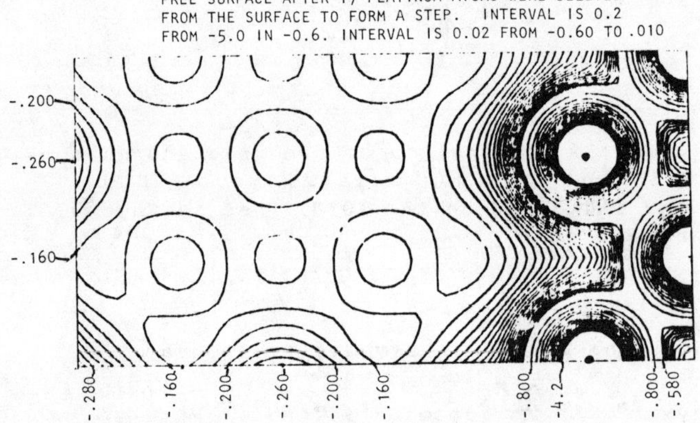

Figure 14. Isopotential contour plot for 17 platinum atoms removed to form a planar microvoid on the (010 surface. Viewing plane cuts through the nuclei of the atoms on the right hand of theis figure. The cores have truncated to facilitate graphing. (after Kennard, Waber & Tsui)

it. Hence the potential is truncated in the spirit of the Pseudopotential method and values inside a sphere of radius $\mathbf{R}_{\mathbf{WS}}$ are set to zero for the purpose of plotting. This enhances one's ability to see the variation in the intercellular region.

In Figure 15, a similar graph is presented for the

potential variation above a pit in the free surface formed by removing 14 atoms. The boundary walls of this microvoid belong to {111}. The hump in the center reduces the electron density

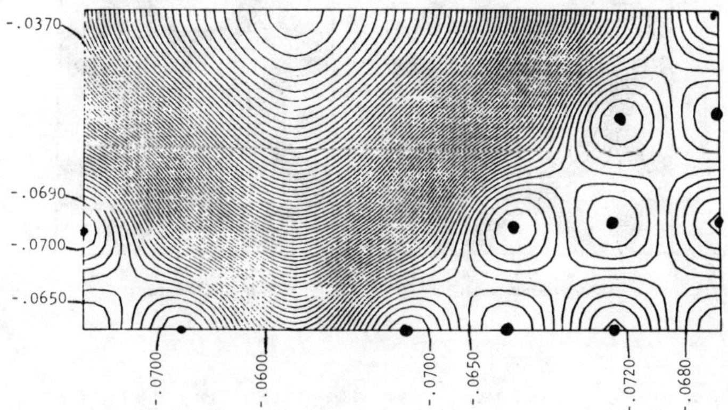

Figure 15. Isopotential Plot for 14 atoms removed from the surface to form a pit bounded by {111} planes. The viewing plane is 1.5 lattice parameters above the free (010)surface

and hence increases the positron lifetime. By counting the contour lines, it can be shown that the hump or maximum potential in the positive direction is higher than in Figure 14. This indicates that the geometry of the microvoid also plays a role in increasing $@_1$ or $@_2$.

Puska and Nieminen (51) have created similar countour plots by superimposing the SCF atomic charge densities, utilizing a method which appears to be that used above; for instance the grid interval is either 1/10 or 1/20 of the lattice parameter. The contour lines of the electronic crystal charge density associated with a monovanancy in BCC or alpha-Fe as well as for a monovacancy in FCC nickel are shown in Figure 15.

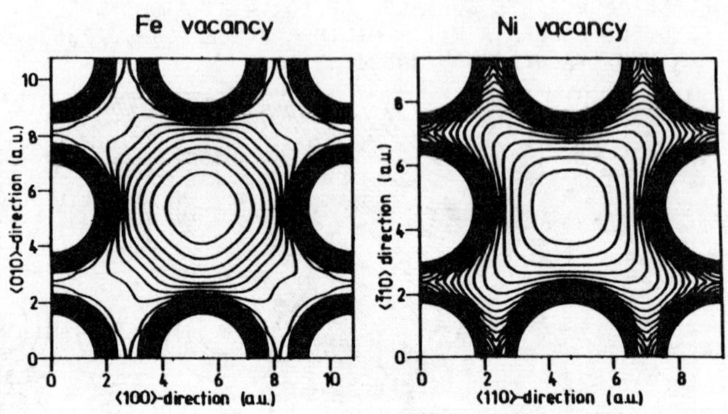

Figure 16. Potential Variation in two lattices in the vicinity of a monovacancy present in iron on the (001) and in FCC nickel on the (111) planes. (after Puska and Nieminen)

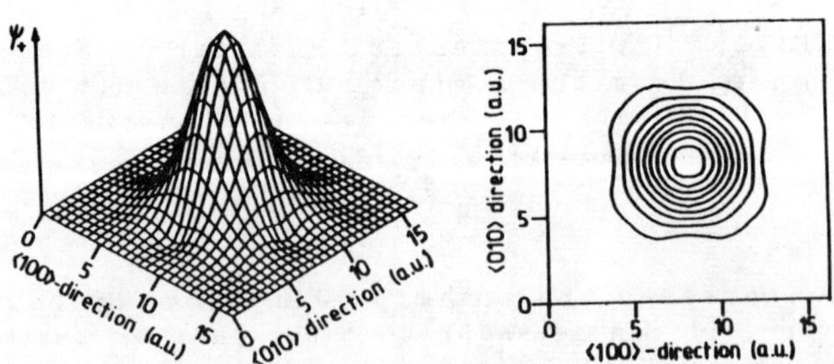

Figure 17. A perspective plot of the spherically symmetric positron wave function distorted by being trapped in monovacancy. (after Puska and Nieminen)

While one edge of the square is labelled as lying parallel to <100>. It is not clear to this reviewer how the arrrangement of atoms can be so similar for both a BCC and a FCC lattice. The viewing plane for iron appears to be (001) and covers four contiguous cells leaving the vacancy in a centered 2x2 structure.

The viewing plane appears to (020) for nickel. Proceeding on the assumption that the illustrations are correct, these authors display the charge density of a positron trapped in such a vacancy for Fe, in Figure 17. For comparison, the electronic charge density associated with a divacancy is shown in Figure 18. The trapped positron is equally distributed over the two vacant sites and has linear symmetry like a homopolar diatomic molecule. They apparently did all of the integrations in a Cartesian coordinate system rather than using the typical multi-center expansions or symmetry operations to simplify the calculations. With both the electronic and positronic charge densities at hand, they calculated the lifetime before annihilation when a positron is trapped in vacancy clusters of increasing size. The lifetimes for iron range from 190 psec for a single vacancy to 304 for six, and they become nearly "saturated" at 386 psec for 15 vacancies in the cluster. At the same time the binding energy increases from 3.0 eV to 7.7 and to 9.7 for the 15 missing atom case. Even though the positron is bound progressively more tightly, the overlap between the valence electrons and positron decreases significantly because the wave function of the single particle is spread over many empty cells. They include the results of similar calculations for Al, Ni and Mo. They all show the similar trends. The author is unaware of any similar plots or calculations for dislocations for any metal.

There is a recent study - which is more accurate than any other of which this reviewer is aware - treating edge dislocations quantum mechanically. Prof. de Hosson (52) has carried it out. The conventional mathematical analysis of dislocations is based on continuum mechanics and since a dislocation is mathematically a pole in the continuum where forces, etc. go to infinity, it is conventional to restrict the expressions to regions outside a core of radius r_c. The defects within r_c i. e., within the dislocation core are not studied. However, de Hosson adopted initially an atomistic approach. He used anisotropic elastic constants to estimate probable positions of atoms relocated by the local force field. In addition, he has summed the interatomic pair-potentials developed by Johnson (1) to get the relaxed atomic

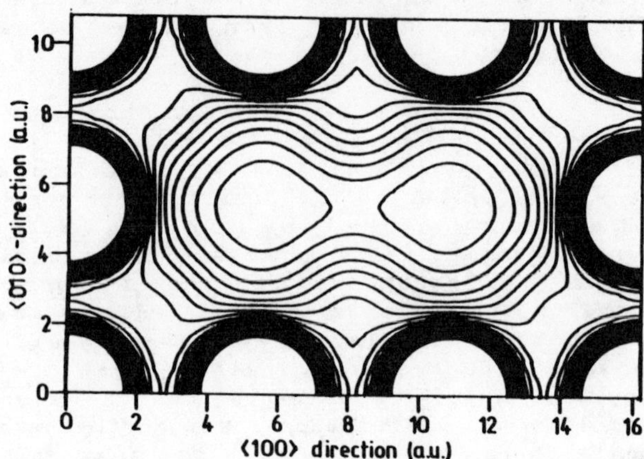

Figure 18. Variation of charge density in the vicinity of a divacancy in iron (after Puska and Nieminen)

positions. The insertion of an edge dislocation lowered the symmetry from O_h for the bulk cubic metal to C_{2h} locally. This reduction in symmetry lifts the degeneracy of various orbitals in the band description of the metal. Some levels move up in energy and some move down in energy. Before showing these graphs it is worthwhile to discuss what would be expected.

On the basis of the dependence of the the various energy bands on the interatomic distance, d_{ij}, Friedel (53) pointed out that the separation between the bottom of the allowed range of energies, E_B and the top E_T would increase as d_{ij} decreases. Further the energy difference between E_B and any given level will increase in the same way, and in specifically, the energy difference $E_F - E_B$ will increase. As suggested in Figure 19, on the compression side, where d_{ij} is smaller, the bottom E_{b1} will be lower than E_B and the top of the occupied levels E_1 will be higher than the value of the Fermi level E_F in the perfect lattice. On the dilational side of the slip plane where the half plane is missing, the spread of energies will be smaller and the top of the occupied levels E_2 will be below E_F and E_{b2} will be above the bulk value E_B. At 0° K, there are now levels occupied above E_F, there must be a local redistribution of the electrons.

Returning to the graphs on iron, in the perfect lattice the 72 electrons per unit cell are distributed over 30 spin polarized states - that is, individual states are assigned for

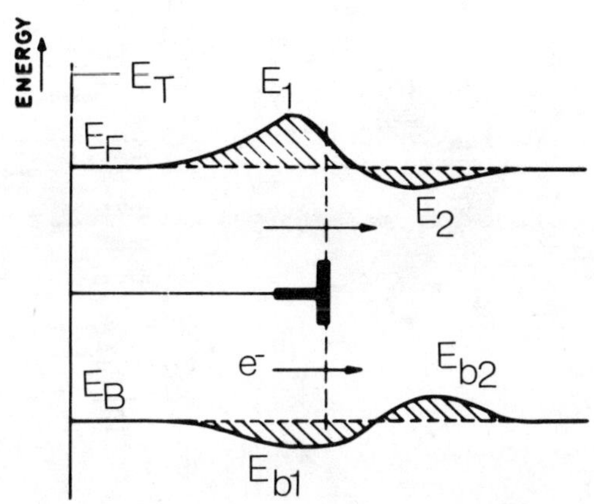

Figure 19. Schematic representation of the energy spread and shifts of occupied levels above and below the slip plane of an edge dislocation. (ater Friedel)

spin up and spin down states - over a range equal to 0.515 eV. Because of the reduction of the symmetry to C_{2h}, the 72 electrons are distributed over 72 states, lying in the range 0.571 eV below the Fermi level. If one were to have considered molybdenum, as de Hosson did, the number of levels would be less since the two spins are usually treated as degenerate in a non-magnetic material.

As indicated above, electrons on the compressional side are "forced out" and repopulated the levels which lie below the Fermi level on the dilated side. About 0.1 electron per atome in the core of $2a_0$ are involved in the creation of the dipole. At a recent Joint U.S. - Japan conference held in Hawaii, a number of authors considered the electronic structure of dislocations using several approximate quantum mechanical methods (54). None of these developed the topic in the depth which de Hosson has done.

(b) Surface Effects The experimental evidence for most of the effects of the medium on the annihilation of thermalized positrons has been presented above. As a note of caution, one

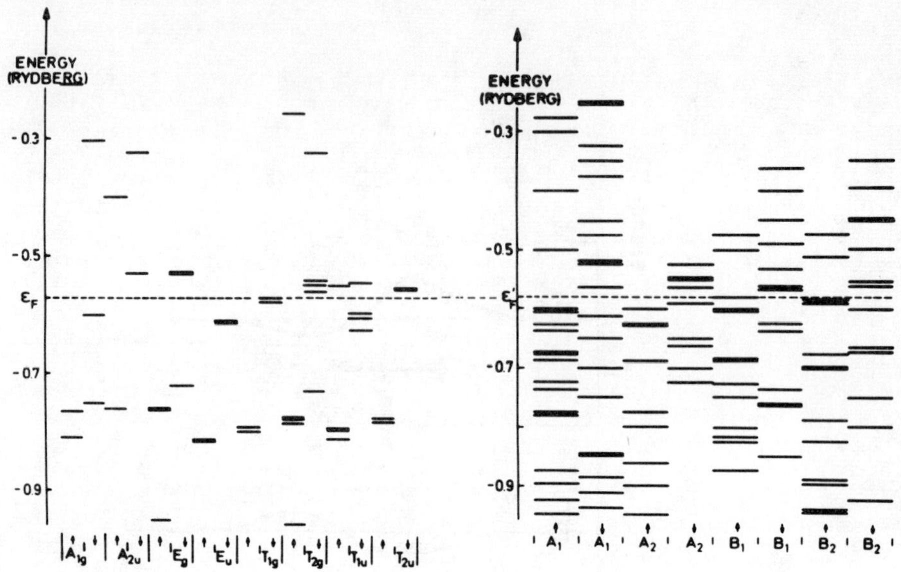

Figure 20. The distribution of the spin up and spin down molecular levels in a cluster of iron atoms containing a (100) <010> type edge dislocation. The left hand panel is for the perfect lattice and the right hand for the dislocated arrangement. (after de Hosson) type

fact seems to have been overlooked by those who do experiments on metals, and that is that almost every one of the technological significant metals are covered with a thin oxide layer. Although it may be only a few angstroms thick when metals are allowed to sit at room temperature in the atmosphere, this film will grow in thickness to a few tenths of the wavelength of light (to ca. 500 A^o) when heated to moderate temperatures, say 600° K. Many experiments have been made, for example, on the effect of temperature on Doppler broadening without great attention being paid to exposure of the sample to the low pressure gas present even if the runs are conducted in an evacuated system. This author is not aware of specimens being brought back down to near room temperature and reheated to look for hysteresis effect or any changes induced during the high temperature exposure.

It is common practice to measure S, the shape parameter of a metal as a function of the temperature and to use the steeply sloping portion of the S vs. T curve to estimate the equilibrium concentration of thermally generated vacancies. The heats of formation derived from this assumption are in good agreement with other methods. But seldom are the specimens brought back to other portions of the curve to establish whether the equilibrium concentration can be re-established under new conditions, i.e., will the excess vacancies anneal out?

Surface effects could be important, but we know little about the effect of progressive oxidation on the positron annihilation processes. The only study which comes to mind is is in the pioneering work by Lynn (55) who exposed oxide free aluminum to a controlled number of Langmuirs of oxygen at a variety of temperatures. A large fraction of the positrons appear as Positronium. He was able to demonstrate that the amorphous oxide initially formed, readily crystallized on the surface at modest temperature considering the Al_2O_3 is a refractory compond. A number of effects are observed, which are incompletely understood. There is a considerable amount of information on the interaction of positrons with surfaces in a recent review (56).

There are a variety of growth laws for oxide films, logarithmic, cubic, inverse logarithmic which are experimentally observed at temperatures in the range of 300 to 700° K and these growth laws are connected with the development of space charges and high electrostatic fields in the oxide layer; these strongly modify the diffusion of cations and/or anions (which are the charge carriers) through the film and limit the thickness achieved under those ambient conditions. The "usual" growth laws are the parabolic and the linear laws. In the first of these growth rates, the thickness y is given by the expression $dy/dt = D/y$, i.e. is diffusion controlled. In the latter $y = \text{const.}$ and is attributed to an oxide product film breaking away from or not adequately covering the metal so that the reaction is interface controlled. These two cover many of the cases but by no means all of them. The other growth laws have been identified with significant departures from either of the two idealized cases.. These uncommon cases, namely, cubic, logarithmic, etc. specifically apply to the behavior of thin oxide films. But it is impractical to discuss this subject in the necessary detail in this review.

It is sufficient to say most metals when used or studied are oxide-covered, so we need to pay a little attention to the effect of the ionic compound formed. There is a surface-source correction frequently of unkown magnitude which "infects" much experimental data. High energy positrons will readily penetrate such a film and sample the bulk of the crystal. Still some will be stopped at or drift back to the metal-oxide film. What are their charcteristics? We need to know so that the question can be dispensed with once and for all.

Such thin flms may not be technologically important. and eve though they may even turn out to have trivial effects on positron rather than positronium annihilation, they offer another tool for probing the effect of surface defects on the behavior of materials. To illustrate a few, fracture is initiated at surfaces, corrosion, erosion and wear are surface dominated. Positron annihilation is non-destructive and very specific. A detailed study of such surface-oriented problems could prove illuminating and rewarding.

Already there are a variety of techniques such as the transmission electron microscope which are good for studying the effect of dislocations and dislocation interactions in a metal. Measuring the unique near-the-surface region of a metal that may be the unique, and in the long run, the significant non-destructive probing ability of positrons.

SUMMARY

In summary, a range of information has been presented on the use of positron annihilation in studying defects in metals and ionic compounds. The major emphasis has been on lifetimes and comparatively little attention has been devoted to angular correlation, which gives a considerable detailed information about the momentum of the electrons, or to Doppler Broadening which is frequently used to detect monovacancy formation in metals and alloys. The latter method is faster than lifetime measurements whereas the angular correlation is significatly slower than either of the other two. It is because the lifetime is directly related to the local density of electrons in the vicinity of the defect and hence can readily be interpreted, that this reviewer has given greater coverage to this type of measurement. The other two types of positron work have unique advantages, and the reader is urged to seek other sources if the type of information he needs is more easily found with them.

REFERENCES

1. R. A. Johnson, *Phys. Rev.*, 134 (1964) 1329 and *Phys. Rev.*, 145 (1966) p. 423.

2. J. T. M. de Hosson, *"Interatomic Potentials and Crystalline Defects"*, J. K. Lee, ed. Pittsburgh Conf. (AIME, Warrendale Pa, 1980)

3. A. Bisi, A. Fiorentini and L. Zappa, *Phys. Rev.*,

131 (1963) p. 1023.

4. W. Brandt, H. F. Wuang and P. W. Levy, _Proc. Int'l Symp. on Colors Centers in Alkali Halides_ p. 46 (Rome,1968)

5. A. Dupasquier, "Positrons in Solids": p 156 in _Topics Curr. Phys._ P. Hautojarvi, ed: **12** (1979).

6. C. Bussolati, A. Dupasquier and L. Zappa, _Il. Nuovo Cimento_ B52 (1967) p. 529.

7. A. B Kunz and J. T. Waber, _Solid State Comm.,_ 39 (1981) p. 861.

8. M. Bertolaccini, A. Bisi, G. Gambarini and L. Zappa, _J. Phys. C., Solid State Phys._ 4 (1971) p. 734.

9. M. Bertolancini and A. Duspasquier, _Phys. Rev.,_ **B**2 (1970) p. 8962

10. James T Waber, C. L. Snead, Jr., "Low Temperature Positron Lifetime and Doppler Broadeneing Experiments For Single Crystal Nickel Oxide Containing Catio Vacancies" Paper M-7 in _Sixth Intl Conf. on Positron Annihilation"_ in Arlington April 1982, P. G. Coleman, S. C. Sharma and L. M. Diana, eds. (North Holland Publ., New York, 1983)

11. N. L. Peterson and W. K. Chen, "Effect of Impurity Content on Cation Vacancy Diffusion and Correlation in Ionic Solids," in _Mat. Sci. Rev.,_ 9 (1975) pp. 41-55.

12. J. D. Christian and W. P. Gulbraith, _Oxid. Metals,_ **(1975)** p. 1.

13. J. Novotny and A. Sadowsky, _Bull. Akad. Pol. Sci., Ser. Sci. Chem._ 25 (1977) p. 825 (in English).

14. I.K. McKenzie and J. Fabian, _Il Nuovo Cim._ 55 (1980) p. 162..

15. A. B. Kunz and J. T. Waber, "Concerning the Trapping of Positrons in Ionic Solids", Paper P-146, in _Proc. VIth Int'l Conf. Positron Annihilation_ (Arlington, 1982) eds and publisher as in ref.12.

16. A. B. Kunz and M. B. Surrat, _Phys. Rv._ B19 (1979) p. 2352.

17. A. Duspasquier, "Positronium-Like Systems in Solids"

paper presented at Varrena Conf, Intl. School Of Physics, Varrena, Italy, Sept. 1981

18. A. Held and S. Kahana, Can. Jour. Phys., 42 (1964) p.19.

19. A. C. Switendick, J. Less Comm. Metals, 74 (1) (1980) p. 189.

20. J. N. Pratt and R. G. R. Sellors, "Electrotransport in Metals and Alloys," P. 167, in Diffusion and Defect Monographs, Y. Adda, A. D. Le Claire, L. M. Shifkin and F. N. Wohibier, eds.: Trans. Tech S. A., Richer, Switzerland, 1973.

21. C. L. Jensen and D. E. Fields, Scripta Met., 15 (1981) p. 101. For a discussion of a possible error, see H. Wipf, ibid p. 733.

22. A. P. Mills, "Experimentation with Low Energy Positron Beams", presented at Int. School of Physics (Varenna, 1981).

23. See various papers in "Hydrogen Embrittlement" M. Bernstein and A. J. Thompson, eds. AIME. Warrendale Pa. 1982.

24. W. Brandt, G. Coussot, R. Paulin, Phys. Rev., Lett., 23 (1969) p. 522.

25. A. Greenberger, A. P. Mills, A. Thompson and S. Berko, Phys. Rev. Lett., 32A (1970) p. 72.

26. T. Hyodo and Y.Takakuso, J.Phys.Soc.Japan, 49 (1980) p. 2243 and p. 2248. See also ibid 42 (1977) p. 1065 and 45 (1978) p. 795.

27. P. Manuel, Paper presented at Intl Schol of Physics (Varenna, 1981).

28. C.H. Hodges, B.T. A. McKee, W. Trifthauser and A. T. Stewart, Canad. Jour. Phys., 50 (1972) p. 103.

29. G. Gambarini and L. Zappa, Phys Lett., 35A (1971) p.193.

30. K. G. Lynn and P. Granatelli, "Defect Characterization with Positron Annihilation," p. 169 in Non-Destructive Evaluation: Microstructural Characterization and Reliability Strategies, O. Buck and S. W. Wolfe, eds.: AIME Warrendale, Pa 1981)

31. P. Hautojarvi A. Verhanen and V. S. Mikhalen, Appl. Phys., ii (1976) p. 191

32. John Hirth, "Effects of Hydrogen on the Properties of Iron and Steel", in Met. Trans., 11A (1980) p. 861.

33. L. C. Smedskjaer, M Manninen and M. J. Fluss, J. Phys., F Met. Phys., 10 (1980) p. 2237

34. H. E. Hjelmroth, J. G. Rasmussen and J. Treff, Ch. Nuovo Cim., 41B (1977) p. 284.

35. P. Hautojarvi, K. Johnson, P. Verhanen and Yli-Kauppilla, Phys. Rev. Lettr., 44 (1980) p. 1326

36. P. Cotterill and R. R. Mould, pp. 42ff in "Recrystallization and Grain Growth in Metals, Wiley, New York, 1976.

37. Vandermeer and Gordon, quoted as ref. 185 in present ref. 36.

38. McLean quoted as ref. 15 in present ref. 36.

39. H.-E. Schaefer, K. Maier, M. Weller, D. Herlock, A. Seeger and J. Diehl, Scripta Met., 11 (1977) p. 803.

40. A. Sato and M. Meshii, Phys. Solidii Stat., 22 (1974) p. 253.

41. R. A. Johnson, Phys. Rev., 134 (1964) p. A1329.

42. K. Hinoda, S. Tanigawa and M. Doyama, J. Phys. Soc. Japan, 39 (1975) p. 545 and ibid 41 (1976) p. 2037

43. J. Diehl, U. Merbold and M. Weller, Script Met., 11 (1977) p. 811.

44. D. Seegers, L. de Schlepper, M. Dorikens, L. Dorikens-Vanpraet, G. Kruyt, L. Stals and P. Moser, "Isothermal Annealing and Detrapping from Dislocation-Like Defects in the Prevacancy Region of Iron", Phys. Lettr. (in press)

45. D. Seegers, F. Van Brabander, L. Dorikens-Vanpraet, "Isothermal Annealing of Room Temperature Deformed Iron Studied with Positron Annihilation Techniques" - Paper P-20 in Arlington Conf.on Positron Annihilation" Arlington. TX

46. B. Pagh, H. E. Hansen, B. Nielsen, G. Trumpy and K. Petersen,"Temperature Dependence of Positron Annihilation Parameters in Neutron Irradiated Mo" Paper M-5, at Arlington Conference.

47. B. Nielsen and K. Petersen, "The Positron Response form Dislocations in Deformed Mo" Paper P-15, at Arlington Conference.

48. B. Nielsen, A, van Ween, L. M. Caspers, W. Lourens, G. Trumpy and K. Petersen, "Interactions between Hydrogen and Defects in Metals Studied by Positron Annihilation Technique", Paper P-37, at Arlington Conference.

49. James T. Waber, E. B. Kennard and Yu-ping Tsui, "Surface Energy and Surface Potentials of Transition Metals" p. C3 in J. de Phys., Colloque C3, Suppl. au Nr. 5-6, Tome, 33 (June 1972).

50. David Liberman, James T. Waber and Donald Cromer, Phys. Rev., 137 (1965) p. A27.

51. M. J. Puska and R. M. Nieminen, "Defect Spectroscopy with Positrons - AGeneral Calculational Method" Research Report 5/1982 University of Jyvaskyla, Jyvaskyla, Finland

52. Jeff Th. M. de Hosson, J. Canad. Phys., 60 (5) (1982) pp. 779-787.

53. Jacques Friedel, p. 365 in "Dislocations", Pergamon Press, New York, N. Y. 1967.

54. See for example, A. Sato and K. Masuda, "Electronic Theory of Screw Dislocation Motion in BCC Transition Metals," and H. Suzuki, "Quantum Mechanical Effects on the Motion of Dislocations in BCC Crystals," in Mechanical Properties of BCC Metals, M. Meshii, ed. AIME, Warrendale, Pa. 1982.

55. K. G. Lynn, Phys. Rev. Lettr., 44 (19) (1980) pp. 1330-1333.

56. K. G. Lynn, "Slow Positrons in the Study of Surface and Near-Surface Defects," presented at Varrena Conf.

NONDESTRUCTIVE EVALUATION OF METALS

BY POSITRON ANNIHILATION TECHNIQUES

S.C. Sharma[*], R.M. Johnson[**], and L.M. Diana[*]

[*]Center for Positron Studies, Department of Physics
[**]Department of Mechanical Engineering
The University of Texas at Arlington, Arlington, Texas 76019

This paper reviews positron annihilation techniques and their utilization for nondestructive evaluation of metals. Some of the results from recent investigations of the annihilation characteristics of positrons in metals with crystalline defects are presented. These results show clearly the high sensitivity of positrons to lattice defects and provide information about the nature and concentration of these defects. The behavior of positrons in metals with crystalline defects is explained by the existence of localized positron states at the site of the defect. The annihilation characteristics of positrons trapped in vacancies, dislocations, and microvoids are significantly different from those of untrapped positrons. Hence, the positrons can be used as a probe to study plastic deformation, fatigue damage, radiation damage, and the like.

Introduction

Applications of positron annihilation techniques to the investigations of a variety of important lattice defects in metals and simple alloys during the last few years have shown a remarkable sensitivity, reasonably good selectivity, and high potential of these techniques for nondestructive examination of materials. It has been amply demonstrated, both experimentally and theoretically, that positron annihilation characteristics are sensitive to changes in the density and momentum distribution of electrons in the region of crystalline defects and that the annihilation gamma rays can provide detailed information about the nature and structure of these defects. Excellent comprehensive reviews of the earlier work in this field are available in the literature.(1)Here we summarize briefly the basic principles of positron annihilation physics and experimental techniques pertinent to nondestructive evaluation and then present results from some selected recent investigations.

Basic Principles of Positron Annihilation

Energetic positrons with average kinetic energies of the order of a few hundred keV are obtained generally from the beta decay of commercially available radionuclides. Having entered a metal, the positrons slow down to thermal energies in a period of about 10^{-12} sec which is about two orders of magnitude smaller than typical values of the mean lifetime against annihilation. The implantation range of positrons injected into a sample depends on the maximum kinetic energy of the positrons and the density of the material according to

$$R^{-1} = \frac{17\ d[\mathrm{gm/cm^3}]}{E_{max}[\mathrm{MeV}]^{1.43}}\ \mathrm{cm}^{-1} \quad , \qquad (1)$$

where R is a range, d is the density, and E_{max} is the maximum energy of the positrons.(2)Typically, the maximum range of positrons of kinetic energies of the order of 0.5 MeV varies from tens to several hundred μm in metallic samples.

In a defect-free single crystal the positron is repelled by the positive ions and is likely to be found in the interstitial regions. The behavior of the positron is described by an extended Bloch state with a probability distribution shown in Figure 1.(3) The annihilation of a positron-electron pair results predominantly in the emission of two gamma rays of energy 0.511 MeV each. The annihilation rate (λ) is given by

$$\lambda = \tau^{-1} = \pi r_0^{\,2} c \int \rho^-(\vec{r})\rho^+(\vec{r})d^3r \quad , \qquad (2)$$

where τ is the mean lifetime against annihilation, r_0 is the classical radius of the electron, c is the speed of light, and ρ^- and ρ^+ are the densities of the electrons and positron, respectively. Thus the annihilation rate of the positron is determined by the density of the electrons at the position of the positron. This fact constitutes the basis for defect studies by positron lifetime experiments, and its usefulness will be discussed in the following sections. In the presence of crystalline defects like vacancies, dislocations, and microvoids, positrons are localized in bound states. This is shown in Figure 2 by the probability distribution

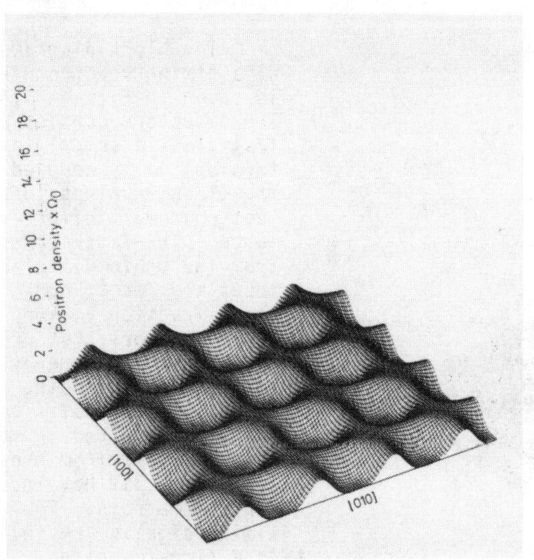

Figure 1 - Positron density x Ω_0 (where Ω_0 is the volume of the primitive unit cell in which the positron density is normalized) in W in the Bloch state of the positron, shown on a network of 25 atomic sites containing no vacancies in the (001) plane. (Courtesy of R.W. Siegel).(3)

of a positron trapped in a vacancy.(3) Thus the characterization of the behavior of a positron changes from an extended Bloch state in a defect-free lattice to a highly localized state in a vacancy. Since the electron density distribution in the region of the crystalline defect is different from that in a defect-free lattice, the trapped positrons annihilate with significantly different characteristics. Positron lifetime measurements provide useful information about the average electron density "sampled" by the positron at the site of the defect and thus constitute a unique method for examining, for example, embryonic vacancy clusters.

The center-of-mass motion of the positron-electron pair produces a broadening in the line-shape of the annihilation gamma rays and a deviation from collinearity in the direction of emission of the two annihilation gamma rays. However, since the positrons are thermalized (average kinetic energy $\sim$ kT), the latter two effects result essentially from contributions due to the motion of the valence and core electrons. This fact constitutes the basis for the usefulness of Doppler-broadening and angular correlation measurements. The shapes of both the angular correlation and Doppler-broadened spectra are modified due to the trapping of positrons at crystalline defects where positrons sample a relatively smaller density of the tightly bound core electrons.

Experimental Techniques

Lifetime Measurements

Positron lifetime measurements are accomplished by making use of the radioactive decay of ^{22}Na and a timing spectrometer. The emission of positrons by this nuclide is accompanied by suitable timing start signals in the form of 1.28 MeV gamma rays from the de-excitation of an excited state of ^{22}Ne. Subsequent annihilation of positrons in the sample results in the emission of 0.511 MeV gamma rays. The detection of one of these annihilation gamma rays provides a stop signal. The positron source is usually prepared by evaporating a few microcuries of commerically available aqueous ^{22}NaCl on a thin metal foil and covering the radioactive spot with the same foil. This source is then sandwiched between two pieces of the sample.

The block diagram of a timing spectrometer is shown in Figure 3.(4) The detectors consist of fast plastic scintillators optically coupled to the glass envelopes of fast photomultiplier tubes. The fast signals from the photomultiplier tubes are fed to constant-fraction-timing discriminators that produce timing and energy signals corresponding to selected ranges of energies absorbed in the scintillators from the 1.28 and 0.511 MeV incident gamma rays. The timing signals are fed to a time-to-pulse-height convertor (TPHC) which produces an output with an amplitude proportional to the time interval between the start and stop inputs. The output of the TPHC is sent to a multichannel analyzer which is gated by the coincidences between the energy selected pulses. The accumulated spectrum is usually described by a sum of exponentials convoluted by the resolution of the spectrometer and a background contribution.

An example of a positron lifetime spectrum and instrumental resolution is shown in Figure 4. The resolution of this spectrometer is characterized by a full-width at half-maximum of about 1.5×10^{-10} sec, which is the best resolution known.(4) In a defect free metal all positrons annihilate from delocalized Bloch states with an annihila-

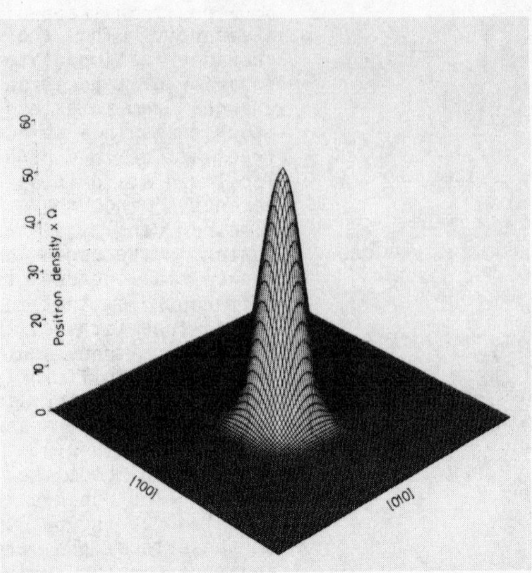

Figure 2 - Positron density x Ω (where Ω is the volume of the supercell in which the positron density is normalized) in W in the vacancy-trapped state of the positron, shown on a network of 25 atomic sites in the (001) plane with the vacancy at the centre. (Courtesy of R.W. Siegel).(3)

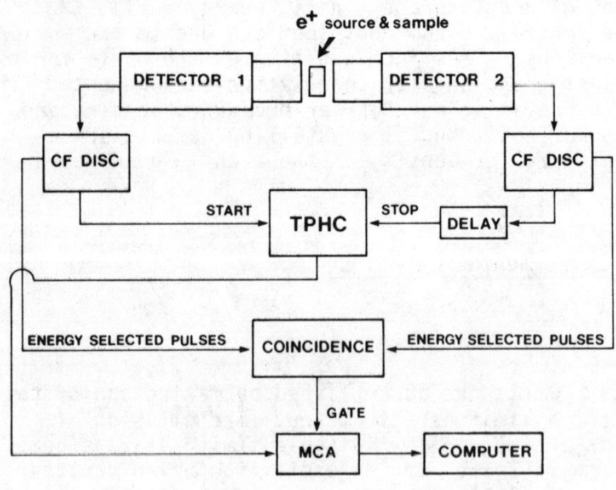

Figure 3 - Schematic diagram of the positron lifetime spectrometer.

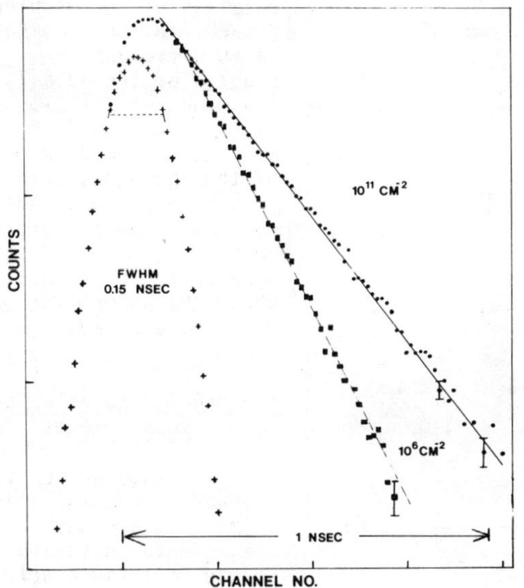

Figure 4 - Typical positron lifetime spectra in annealed and deformed samples. The prompt curve represents resolution of the spectrometer.

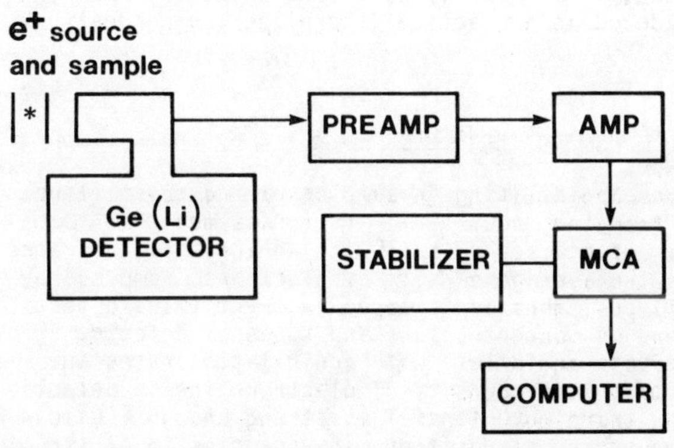

Figure 5 - Schematic diagram of the Doppler-broadening spectrometer.

tion rate that is characteristic of the average electron density in the sample. However, in the presence of crystalline defects in low concentrations positrons annihilate from at least two different states producing a multiexponential spectrum. Such spectra can now be analysed by computers and accurate values of the annihilation rates and relative intensities obtained for the different annihilation modes. High resolution measurements of positron lifetime spectra are essential for a quantitative understanding of the physical processes involved in the behavior of positrons trapped in crystalline defects.

Doppler Broadening Measurements

As mentioned earlier, the motion of the positron-electron pair produces a Doppler shift in the energy of the annihilation gamma ray which can be measured with solid state detectors. A block diagram of the apparatus is shown in Figure 5. A typical Doppler-broadened spectrum and intrumental resolution are shown in Figure 6. As is evidenced from the data of Figure 6 and earlier discussion, the annihilation of positrons with electrons in the region of crystalline defects results in a lineshape that is significantly narrower than that obtained from annihilations from de-localized states. Although the energy resolution of the Doppler broadening spectrometer is insufficient for studies of the details of the momentum distributions, the high counting efficiency and simplicity of the apparatus make this kind of measurement the most promising for nondestructive evaluation, especially in those cases where the underlying processes are known.

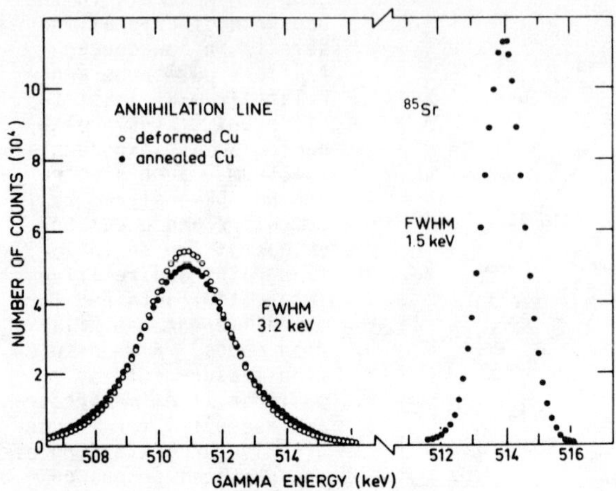

Figure 6 - Doppler-broadened spectra in annealed and deformed copper. The narrowest peak represents approximate energy resolution of the system. All curves have been normalized to equal area. (Courtesy of P. Hautojärvi).(1)

Angular Correlation Measurements

The angular correlation of the annihilation gamma rays depends on the momentum distribution of the positron-electron pairs. These measurements are often used for Fermi surface studies in metals and disordered alloys. The angles involved in these measurements are only a few milliradians which fact demands that the detectors be placed at large distances from the sample (several meters)and that unusually strong radioactive sources be used. Therefore, this is not considered as a practical technique for nondestructive examination.

The Trapping Model

The behavior of positrons annihilating in samples having crystalline defects is explained by the trapping model.(5-7) It is assumed that i) positrons annihilate in a defect-free crystal with an annihilation rate λ_f that is characteristic of the average density of electrons "sampled" by positrons in the crystal, ii) positrons are trapped in crystalline defects with probabilities that depend on concentrations and types of defects, iii) positrons trapped in defects annihilate with annihilation rates λ_D which are lower than λ_f due to a lower density of electrons inside defects. Under the assumption that the trapping potential is strong enough (Calculations show that the binding energy of positrons in vacancies in Al is $\sim$ 2.6 eV(8)) to prevent the escape of the positron from the defect, the time dependence of the number of positrons in the sample is given by (9)

$$\frac{dn_f(t)}{dt} = - \lambda_f n_f(t) - \mu_1 c_1 n_f(t) - \mu_2 c_2 n_f(t)$$

$$\frac{dn_{D1}(t)}{dt} = - \lambda_{D1} n_{D1}(t) + \mu_1 c_1 n_f(t) \tag{3}$$

$$\frac{dn_{D2}(t)}{dt} = - \lambda_{D2} n_{D2}(t) + \mu_2 c_2 n_f(t) \quad .$$

Here n_f, n_{D1}, and n_{D2} are the numbers of positrons in the free (f) and in two defect states D1 and D2. λ_f, λ_{D1}, and λ_{D2} represent annihilation rates in these states; μ_1 and μ_2 are trapping rates per unit concentration of the two types of defects; and c_1 and c_2 are the concentrations of defects. Although these equations can be generalized for any number of defects, there are generally one or two kinds of defects encountered in simple practical cases, for example, monovacancies and divacancies. If no positrons were trapped at time zero, then the solution of these equations provides

$$n_f(t) = n_o \exp[-(\lambda_f + \mu_1 c_1 + \mu_2 c_2)t] \quad ,$$

$$n_{Dj}(t) = \left(\frac{n_o \mu_j c_j}{\lambda_f - \lambda_{Dj} + \sum_{k=1}^{2} \mu_k c_k} \right) \exp[-\lambda_{Dj} t] \, * \tag{4}$$

$$\{1 - \exp[-(\lambda_f - \lambda_{Dj} + \sum_{k=1}^{2} \mu_k c_k)t]\} \quad ,$$

$$\text{for } j = 1 \text{ and } 2 \; .$$

The probability of annihilation between time t and t + dt is then given by

$$P(t)dt = (\lambda_f + \sum_{k=1}^{2} \mu_k c_k)\{1 - \sum_{j=1}^{2} \frac{\mu_j c_j}{(\lambda_f - \lambda_{Dj} + \sum_{k=1}^{2} \mu_k c_k)}\} \, *$$

$$\exp[-(\lambda_f + \sum_{k=1}^{2} \mu_k c_k) t]dt \tag{5}$$

$$+ \sum_{j=1}^{2} \frac{\mu_j c_j \lambda_{Dj}}{(\lambda_f - \lambda_{Dj} + \sum_{k=1}^{2} \mu_k c_k)} \exp[-\lambda_{Dj} . t]dt$$

This probability function indicates that a multicomponent spectrum will be accumulated which consists of exponential terms with annihilation rates and relative intensities given by

$$\lambda_1 = \lambda_f + \sum_{k=1}^{2} \mu_k c_k \ ,$$

$$\lambda_2 = \lambda_{D1} \ ,$$

$$\lambda_3 = \lambda_{D2} \ , \qquad (6)$$

$$I_j = \frac{\mu_j c_j}{(\lambda_f - \lambda_{Dj} + \sum_{k=1}^{2} \mu_k c_k)} \ , \text{ for } j = 2 \text{ and } 3.$$

The annihilation probabilities from these different states are given by

$$P_f = \frac{\lambda_f}{\lambda_f + \sum_{k=1}^{2} \mu_k c_k} \ , \text{ and}$$

$$P_{Dj} = \frac{\mu_j c_j}{\lambda_f + \sum_{k=1}^{2} \mu_k c_k} \ , \qquad (7)$$

Results

Monovacancies to Microvoids

Positron annihilation techniques have been used to investigate monovacancies, and microvoids. The formation energies for vacancies (E_v) have been determined for a number of metals by measuring the temperature dependence of certain characteristics of positron annihilation in these samples. Generally these results are in good agreement with those available from established metallurgical techniques. Here, two examples are presented. The temperature dependencies of the counting rates at the peak of angular correlation curves measured for In, Cd, Zn, Pb, and Al are shown in Figure 7.(10) These data reflect a gradual increase in the probability for trapping of positrons as the equilibrium concentration of vacancies rises with the temperature of the sample. At low temperatures the concentration of vacancies in these samples is too small to trap a significant number of positrons. However, at sufficiently high temperatures the probability of finding a positron in a vacancy approaches 100%. Analyses of these data in terms of the trapping model yield Arrhenius plots shown in Figure 8 and provide precise values for E_v. As a second example, the results from a high resolution lifetime study of positron trapping by vacancies in Pb are presented. Figure 9 shows the temperature dependencies of the two resolved lifetimes and the intensity of the longer-lived component that results from annihilations of positrons trapped in vacancies.(11) In addition to providing precise results for E_v, such high resolution measurements are considered essential to an understanding of the physical processes involved. The high sensitivity of positrons to vacancies and the precision of the derived results are noteworthy. It is known that at temperatures close to the melting point of metals there is a

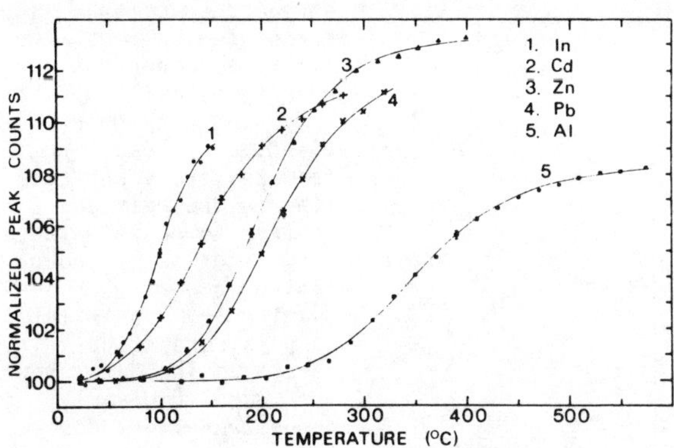

Figure 7 - Counting rates at the peak of the angular correlation curves versus temperature for In, Cd, Zn, Pb, and Al. Normalization assigns the low-temperature limit as equal to 100. (Courtesy of A.T. Stewart).(10)

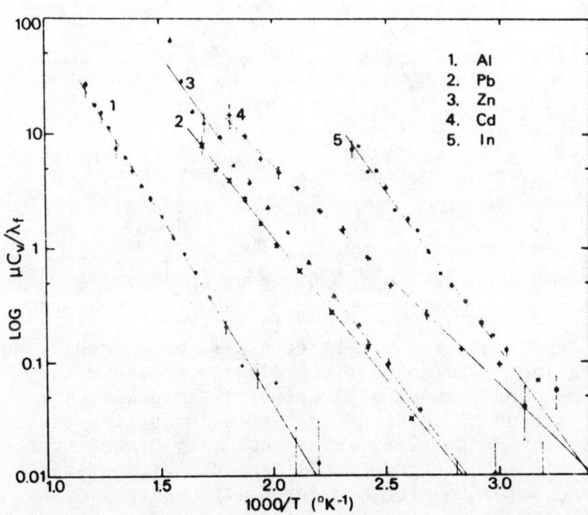

Figure 8 - Arrhenius plots of positron trapping by vacancies for the data shown in Figure 7. (Courtesy of A.T. Stewart).(10)

finite concentration of divacancies. When data from positron annihilation experiments are analyzed by means of the trapping model assuming positron trapping by mono- and divacancies, acceptable results are often obtained for both monovacancy and divacancy parameters.(12) Some of the results obtained for these parameters from positron annihilation experiments are given in Table 1.

Of considerable interest are comparisons among results obtained by positron annihilation, resistivity, and transmission electron microscopy that show high sensitivity and selectivity of positron annihilation experiments for investigating the process of vacancy migration leading to the formation of microvoids. For example, changes in positron annihilation parameters during isochronal annealing of irradiated samples are much more sensitive to the size of the microvoid than are the measurements of resistivity.(15) The results obtained for positron lifetime and angular correlation parameters and resistivity in electron irradiated molybdenum subjected to isochronal annealing are shown in Figure 10.(16) The resistivity data show a major annealing stage at about 260°C where approximately 80% of the radiation induced resistivity is removed. This is followed by annealing stages at about 135 and 675°C where the remaining 20% resistivity is removed. The positron

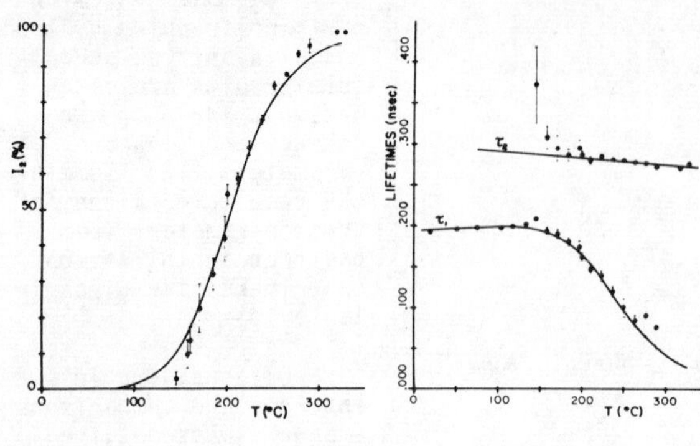

Figure 9 - Lifetimes τ_1 and τ_2 and the intensity of the long lifetime component versus temperature in Pb.(11)

lifetime measurements provide a mean lifetime of 0.118 ± 0.002 nsec in annealed specimens. In irradiated samples, two lifetime components are present below 1000°C, a component with shorter lifetime τ_1 resulting from positrons annihilating in the bulk and another component with a long lifetime τ_2 and relative intensity I_2 resulting from annihilations of positrons trapped in defects. The data shown in Figure 10 clearly demonstrate that positrons are trapped in irradiation-produced defects where they annihilate with a mean lifetime of 0.198 ± 0.005 nsec (about 68% higher than the lifetime in annealed samples) with an intensity of about 65%. Marked changes observed in the values of τ_2 and I_2 above

Table I

Metal	E_{1v}(eV)	E_{2v}(eV)	Reference
Al	0.66 ± 0.09	-	13
Ni	1.74 ± 0.06	3.10 ± 0.23	12
Cu	1.28 ± 0.04	2.26 ± 0.16	12
Zn	0.54 ± 0.02	-	10
Cd	0.39 ± 0.04	-	10
In	0.55 ± 0.02	-	10
Au	0.96 ± 0.02	1.69 ± 0.05	14
Pb	0.58 - 0.65	-	11

200°C indicate an increase in the size of the defects with an accompanying reduction in their concentration. The angular correlation data are consistent with these findings and demonstrate further the migration of vacancies leading to the formation of voids during annealing of irradiated samples. Another example of the sensitivity and selectivity of positron annihilation techniques for microscopic defects is provided by a study of the annealing behavior of vacancies in electron-irradiated α-iron. The results obtained for the lifetime τ_2 and relative intensity I_2 of positrons annihilating from trapped states in electron-irradiated α-iron as a function of isochronal annealing temperature and irradiation dose are shown in Figure 11.(17) Between 77 and 120K the intensity I_2 stays at 100% indicating that all positrons are trapped in defects where they

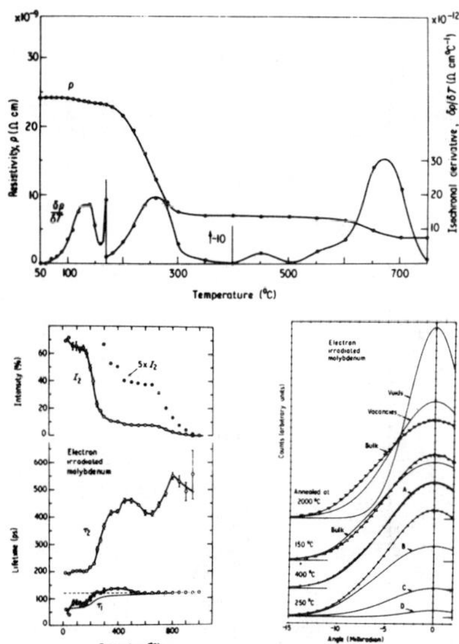

Figure 10 - (a) The recovery of liquid helium resistivity in electron irradiated molybdenum subsequently subjected to isochronal annealing (heating rate 1°C min^{-1}). The derivative $\delta\rho/\delta T$ is also shown to emphasize the annealing stages. (b) Positron lifetime parameters for electron irradiated molybdenum as a function of annealing temperature. The τ_2, I_2 component is due to positrons trapped in vacancies and/or voids. The τ_1 lifetime is for positrons annihilating in the bulk. The lifetime in annealed Mo is shown by a broken line. (c) The measured points and the fitting curves (final fits) of four angular correlation distributions normalized to equal areas (run B). At the upper curve are shown the vacancy and void curves of the same area. Curve A is the fitting curve of the 150°C distribution. Curves B (bulk), C (vacancy), and D (void) are the three components of the 250°C fitting curve. (Courtesy of M. Eldrup).(16)

annihilate with a mean lifetime of about 0.17 nsec. The latter result is known to be the value of the lifetime of positrons trapped in monovacancies in iron. In the sample irradiated with low dose the intensity I_2 decreases to ∿ 75% at 140K, but τ_2 remains constant. This indicates a decrease in the concentration of monovacancies and is consistent with results from electrical resistivity and relaxation measurements. A dramatic change in the annihilation parameters at 230K indicates the formation of vacancy clusters in which positrons sample a much larger empty volume and thereby annihilate with a lifetime that is considerably longer than the value of about 0.17 nsec for positrons trapped in monovacancies. A gradual increase in the value of τ_2 between 235 and 325K indicates an increase in the size of the microvoid.

Dislocations and/or Jogs on Dislocations

A potential of positron annihilation techniques for studying dislocations had been realized almost fifteen years ago as a result of an observation of a significant narrowing of the angular distribution of annihilation gamma rays in plastically deformed aluminum.(18) Since then a number of investigations have been made in order to study the dependencies of positron annihilation parameters on the density of dislocations. Due to well known difficulties in the characterization and isolation of dislocations and in obtaining accurate densities of dislocations, most of the early investigations were limited to obtaining qualitative data on the behavior of positron annihilation parameters in deformed samples. Nevertheless, they demonstrated the high sensitivity of positron annihilation parameters to crystalline defects produced during deformation.(1) High-precision and high-resolution positron annihilation experiments are now

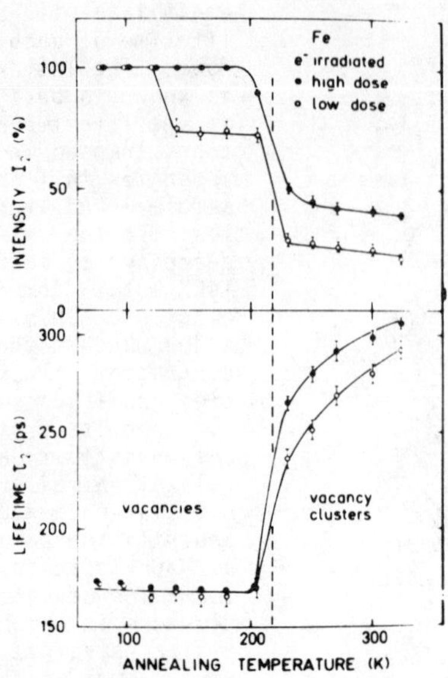

Figure 11 - The lifetime τ_2 and intensity I_2 of trapped positrons in electron-irradiated high-purity -iron as a function of isochronal annealing temperature. The high and low irradiation doses correspond to resistivity increases of 560 nΩcm and 220 nΩcm, respectively. (Courtesy of P. Hautojärvi).(17)

becoming available in carefully prepared samples that are well-characterized in terms of the type and concentration of dislocations. Results from these experiments will be extremely useful to our understanding of the behavior of positrons in deformed samples. Specifically, these experiments are expected to provide accurate results that will help us understand in some detail the nature of the trapping and annihilation of positrons at dislocations and/or jogs on dislocations and the sensitivity of positrons to edge and screw dislocations.

Figure 12 shows the observed dependence of a shape parameter derived from the Doppler-broadened spectra of annihilation gamma rays as a function of the density of dislocations in copper single crystals.(19) In this study the single crystals were oriented and bent so as to introduce primarily edge dislocations ranging in density from 4×10^5 to 1.6×10^8 cm^{-2}. The densities of dislocations were calculated by using the radius of curvature of the bend and the orientation of the preferred slip plane relative to the top surface of the sample. For a range of densities of dislocations from $(0.5\text{-}1.6) \times 10^8$ cm^{-2}, this shape factor varies almost linearly with the dislocation density. Similar results have been obtained for the trapping of positrons by edge dislocations in zinc single crystals. An analysis of these data following the trapping model provides a value of $(1.6 \pm 0.4) \times 10^{11}$ sec^{-1} for the trapping rate per unit density of dislocations, about 100 times higher than the corresponding value obtained for the trapping of positrons by vacancies.(20)

Preliminary results from a recent study of the trapping of positrons by dislocations in high purity copper single crystals are shown in Figure 13.(21) These data were measured simultaneously by high resolution (FWHM $\sim$ 0.15 nsec) positron lifetime and Doppler broadening spectrometers with the goal of obtaining accurate information about the decay rates and intensities of the different exponential components that can now be resolved in the measured spectra. The densities of dislocations in the single crystals were introduced by straining uniaxially. The densities were determined from etch-pit counts and also calculated by utilizing computed values of resolved shear stress. It should be noted that the dislocation densities can be calculated with some reasonable accuracy for a wide range of densities, the measured positron lifetime spectra can be

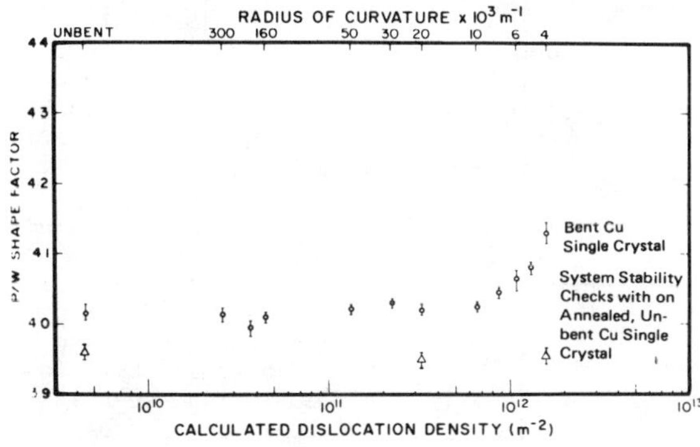

Figure 12 - The circular data points represent experimental Peak to Wings shape factor vs dislocation density calculated from the radius of curvature of the bent Cu single crystal. The triangular data points represent only system stability checks with an annealed, unbent Cu single crystal.(19)

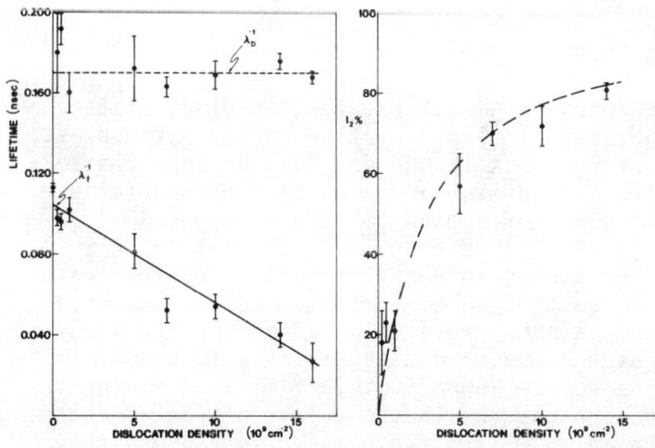

Figure 13 - Positron lifetimes and intensity of the long lifetime component versus the density of dislocations in copper single crystals.(21)

resolved with reasonable confidence into lifetime components due to annihilations from different positron states, and the annihilation parameters derived from two different experiments are consistent. Results from detailed analyses of these data are expected to provide important information about the role played by dislocations in the trapping and subsequent annihilation of positrons, apparent differences in the trapping of positrons by edge and screw dislocations, and the mechanism of the trapping of positrons by crystalline defects.

Fatigue Damage

Positron annihilation techniques have been used to study fatigue damage with limited success because of the complexity of the nature of defects produced during deformation. Nevertheless, the ability of the positron annihilation techniques to monitor changes in the gross features of the sample subjected to fatigue damage, irrespective of microscopic details, has been demonstrated. Figure 14 shows the observed variation in positron lifetime as a function of the number of fatigue cycles applied to soft and hard samples of 4340 steel that were subjected to a maximum stress equal to two-thirds of the yield stress. In the soft sample the positron lifetime increased rapidly, as a result of fatigue hardening, from an initial value of about 0.12 nsec to a value of

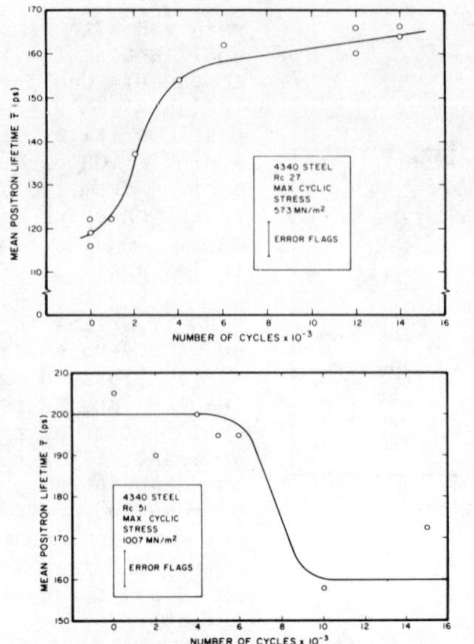

Figure 14 - Mean positron lifetime versus number of fatigue cycles for 4340 steel of initial hardness values of RC-27 and RC-51.(22)

about 0.17 nsec at fracture. In the hard sample of 4340 steel the positron lifetime decreased by almost 18% at about 10^4 fatigue cycles. These results have been interpreted in terms of work softening and/or migration of interstitial impurities to positron trapping sites.(22) The usefulness of positron annihilation techniques for design and for failure prediction is indicated by a long region of small but significant slope seen in the behavior of the positron lifetimes as a function of fatigue cycles.

The measurements of positron lifetime and Doppler-broadening in a sample of 304 stainless steel subjected to cold rolling and fatigue cycling have shown the production of dislocations and vacancy clusters in the initial stages of fatigue damage.(23) Marked changes in the annihilation parameters are observed as a function of the magnitude of the cold work and the number of fatigue cycles. Carefully designed experiments performed under controlled conditions are needed to develop understanding and methods for the ready utilization of positron annihilation techniques for quantitative nondestructive examination.

Plastic Zone of a Propagating Crack

An attempt has been made recently to utilize the technique of the Doppler broadening of positron annihilation radiation to study the plastic zone associated with a propagating crack in copper.(24) An approximately 17 mm long crack was introduced into a standard compact tension (fatigue) sample machined from a 3.18 mm plate of commerically pure copper by tension-tension cyclic loading. A plastic zone ahead of this crack was developed by applying a range of stress intensities (with a maximum value of about 1.76×10^7 Pa $\sqrt{m}$) corresponding to a minimum to maximum load range, r, equal to 0.1. Figure 15 shows results obtained for the lineshape parameter of Doppler broadened spectra measured as a function of relative displacement of a ^{22}Na source approximately 1 mm in diameter. These data clearly show marked enhancement in the values of the lineshape parameter for positrons annihilating within the plastic zone where the density of dislocations is expected to be much higher than that in regions sufficiently far from the crack tip. The full-width at half-maximum of these data ($\sim$ 8 mm) agrees reasonably well with a value of the diameter of the plastic zone estimated by using known values of the stress intensity factor and the yield strength of the sample. Figure 16 shows similar data after the crack in the same sample was developed to a length of about 25 mm.

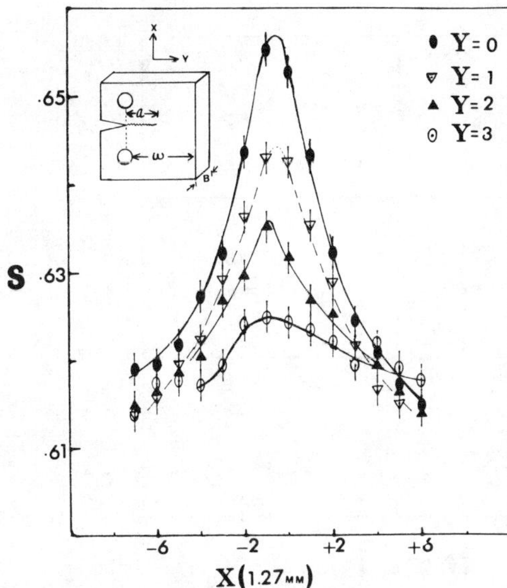

Figure 15 - The S parameter versus relative displacement of the source of positrons in the vicinity of a plastic zone on the fatigue cracked sample. The abscissa corresponds to displacement of the source in a line perpendicular to the length of the crack. The values of Y indicate the number of steps in units of ∿ 1.27 mm by which the source was displaced along the length of the crack prior to making measurements in X direction. X = Y = 0 represents visually estimated tip of the crack. The curves through data points represent hand-drawn smoothed curves to aid the eyes. (24)

These data show unmistakably an extended plastic zone as a result of the application of additional fatigue cycling to the sample. Irrespective of the knowledge about the detailed nature of the processes of trapping and annihilation of positrons in the complex defect structure expected in the plastic zone, these data indicate the high potential of positron annihilation techniques for non-destructive examination of flaws in crystalline solids.

Concluding Remarks

The results presented in this article show the high sensitivity and reasonably good selectivity of positrons to a variety of crystalline defects and demonstrate the usefulness of positron annihilation techniques for nondestructive evaluation of materials. High resolution positron lifetime and Doppler broadening spectra should be investigated simultaneously in carefully prepared and characterized samples in order to obtain a clear understanding of the physical processes involved in the trapping and annihilation of positrons in defect states. Successful utilization of positron annihilation techniques for solving NDE related problems appears possible in the near future.

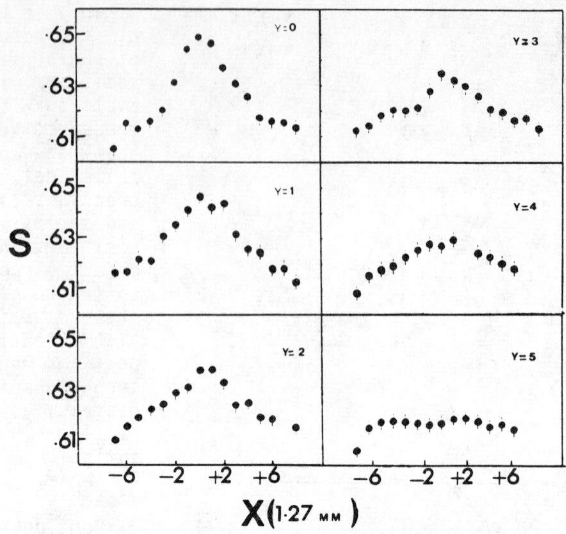

Figure 16 - The S parameter versus relative displacement of the source of positrons in the vicinity of an extended plastic zone on the same fatigue cracked sample as that in Figure 15. The X values correspond to displacement of the source in a line perpendicular to the length of the crack. The values of Y indicate the number of steps in units of $\sim$ 1.27 mm by which the source was displaced at right angles to the X direction prior to measurements along X direction. X = Y = 0 represents the new location of the visually estimated tip of the crack. (24)

References

1. P. Hautojärvi, ed., Positrons in Solids, Topics in current Phys, Springer-Verlag, New York (1979) and references therein.

2. W. Brandt, Appl. Phys., 5 (1974) pp. 1-23.

3. R. P. Gupta and R. W. Siegel, J. Phys., F 10 (1980) pp. L7-13.

4. S. C. Sharma, "A High Resolution Positron Lifetime Spectrometer" (to be published).

5. W. Brandt, in Positron Annihilation, A. T. Stewart and L. O. Roellig, eds.; Academic Press, New York, 1967, pp. 155-182.

6. D. C. Connors and R. N. West, Phys. Lett., 30A (1969) pp. 24-25.

7. B. Bergersen and M. J. Stott, Solid St. Comm., 7 (1969) pp. 1203-1205.

8. J. Arponen, P. Hautojärvi, R. Nieminen, and E. Pajanne, J. Phys. F, 3 (1973) pp. 2092-2108.

9. S. C. Sharma, "High-Resolution Lifetime Study of Positron Trapping in Pb and Pb-4 at. % Cd," Ph.D. Dissertation, Brandeis University, 1976 (unpublished).

10. B. T. A. McKee, W. Triftshäuser, and A. T. Stewart, Phys. Rev. Lett., 28 (1972) pp. 358-360.

11. S. C. Sharma, S. Berko, and W. K. Warburton, Phys. Lett., 58A (1976) pp. 405-408.

12. S. Nanao, K. Kuribayashi, S. Tanigawa, and M. Doyama, J. Phys. F, 7 (1977) pp. 1403-1419.

13. M. J. Fluss, L. C. Smedskjaer, M. K. Chason, D. G. Legnini, and R. W. Siegel, Phys. Rev. B, 17 (1978) pp. 3444-3454.

14. G. Dlukek, O. Brummer, and N. Meyendorf, Appl. Phys., 13 (1977) pp. 67-70.

15. P. Hautojärvi, J. Heinio, M. Manninen, and R. Nieminen, Phil. Mag., 35 (1977) pp. 973-981.

16. M. Eldrup, O. E. Mogensen, and J. H. Evans, J. Phys. F, 6 (1976) pp. 499-521.

17. P. Hautojärvi, T. Judin, A. Vehanen, J. Yli Kauppila, J. Johansson, J. Verdone, and P. Moser, Solid St. Comm., 29 (1979) pp. 855-858.

18. S. Berko and J. C. Erskine, Phys. Rev. Lett., 19 (1967) pp. 307-309.

19. M. L. Johnson, S. F. Saterlie, and J. G. Byrne, Met. Trans. A, 9 (1978) pp. 841-845.

20. J. D. McGervey and W. Triftshauser, Phys. Lett. A, 44 (1973) pp. 53-54.

21. S. C. Sharma, Y. J. Attaiyan, R. M. Johnson, L. M. Diana, S. Y. Chuang, and P. G. Coleman, "High-Resolution Positron Lifetime and Doppler Broadening Measurements of Positron Trapping at Dislocations in Copper," paper presented at the Sixth International Conference on Positron Annihilation, Arlington, Texas, April, 1982.

22. K. G. Lynn and J. G. Byrne, Met. Trans A, 7 (1976) pp. 604-606.

23. K. Nishiwalei, N. Owada, K. Hinode, S. Tanijawa, K. Shibata, T. Fujita, and M. Doyama, "The Study of Fatigned Stainless Steel by Positron Annihilation," Proceedings of the Fifth International Conference on Positron Annihilation, Japan, 1979, 9A-III-4, pp. 177-180.

24. S. C. Sharma, R. M. Johnson, Y. J. Attaiyan, L. M. Diana, and P. G. Coleman, "Doppler Broadening of Positron Annihilation Radiation Around Fatigue Cracks in Copper," paper presented at the Sixth International Conference on Positron Annihilation, Arlington, Texas, April, 1982; S. C. Sharma, Y. J. Attaiyan, R. M. Johnson, and L. M. Diana, "A Study of Dislocations in Copper Single Crystals and Around Fatigue Cracks By Doppler Broadening of Positron Annihilation Radiation," submitted for publication in Met. TRans. A., May, 1982.

DETECTION OF DEFECTS AND METALLURGICAL VARIATIONS IN METAL SURFACES

H. S. Isaacs

Brookhaven National Laboratory
Upton, New York 11973
USA

A highly sensitive non-destructive method employing a scanning technique is described for sensing variations in metal surfaces in aqueous environments, using AC impedance. The probe consisted of a dual compartment with two orifices passing through its base and housing a pseudo-reference electrode and an electrode for current injection. Measurements have shown that the technique can detect differences due to metallurgical variations in welds and its heat affected zones and the presence of cracks in pre-cracked specimens. Details of the methodology and principles are given.

Introduction

There are very few non-destructive techniques which resolve surface heterogeneities on surfaces when exposed to aqueous environments. The differences in surface properties can influence the corrosion behavior of bare metal surfaces, the deposition and adhesion of electrodeposits and the corrosion resistance of coatings, as well as the electrochemical behavior of electrodes used in fuel cells or for water electrolysis. The variations in surfaces may arise because of surface preparation, the deposition of contaminants or metallurgical structures.

DC scanning methods are available for the study of surfaces when relatively high current densities flow (1,2), but in many cases dissolution leads to permanent surface changes and cannot be considered to be a non-destructive technique. To overcome these difficulties, an AC technique was developed (3). The advantages of the AC over the DC techniques are that only very small currents are required, the amplification of the signals can be easily achieved with available electronic instrumentation, and extraneous noise can be filtered out. Because of the small AC currents required, there are no surfaces changes due to the measuring system and the technique can be considered non-destructive.

The AC technique will be described in this paper along with the basic principles and instrumentation. Examples of the application and sensitivity of the technique will be described for cracked and welded specimens.

Principles of the AC Scanning Technique

Impedance measurements have been used extensively to study the metal/solution interface (4,5). The dominant process at the surface depends both on the signal frequency and the metal studied. At frequencies above 100 kHz, the double layer dominates the impedance, giving mainly a capacitative component with metals having no oxide or when the oxide has a high conductivity. With metals covered by a high resistance oxide as for example, aluminum, the capacitance is related to the thickness of the non-porous oxide (6). At lower frequencies, the reactions taking place on the metal become the major contributors to the impedance, and at very low frequencies, the resistance of the interface is equal to the slope of the polarization curve, or polarization resistance. Using DC techniques, this resistance has been shown to be related to the corrosion rate (7). At intermediate frequencies, the impedance is usually related to the kinetics determined by diffusion or adsorption processes. When high resistance surface oxide layers or coatings are present, the response is related to their relaxation processes or resistivity variations and is sensitive to physical defects or pores(6).

All the factors which determine the impedance depend on the nature of the surface. For example, the double layer properties will vary with changes in the metallurgical structure of the surface, and this variance is measured when the surface is scanned. Similarly, the electrochemical reactions or corrosion rates will depend on the underlying metal and these variations are also sensed by the impedance difference. In the case of defects in oxides or coatings, the low resistance paths across the surface layer are easily found using the scanning AC method(3).

AC measurements require the use of a three electrode cell, incorporating (i) the specimen surface being studied, (ii) a counter or auxiliary electrode to inject the AC current into the solution and across the specimen's surface, and (iii) an electrode to sense the change in the potential resulting from the AC current. This cell also has the basic configuration used in DC electrochemical or corrosion measurements. Two electrode cells have been used to measure the impedance, but these cells require large counter electrodes with very much lower impedance than the surface being studied. The three electrode cell overcomes this difficulty as the measurements are independent of both the counter and the reference electrode impedances. In addition, for the AC scanning application, the smaller the effective size of the counter and reference electrodes, the greater will be the spacial resolution. It should be noted that the effective size of these electrodes is actually determined by their solution paths entering the measuring cell and not by the physical dimensions. The reference electrode used is not a true reference, as the DC potential is not important when measuring the AC response.

An example of an AC scanning electrode probe is shown in Figure 1. The probe contains two compartments with small orifices passing through its base. The base is scanned close to and parallel with the surface being studied. These orifices form the solution path between the reference and counter electrodes of a thin layer cell restricted between the base of the probe and the surface of the sample. The advantage of having a thin layer cell arise because the solution resistance path between the base of the probe and the surface to the bulk solution is high. The high resistance restricts the flow of current into the bulk solution and beyond the probe and improves the spacial resolution.

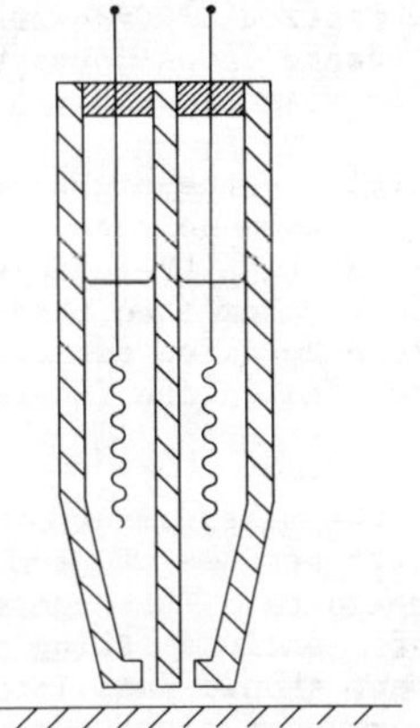

Figure 1. Schematic of the scanning probe showing two compartments for the counter and pseudo-reference electrodes.

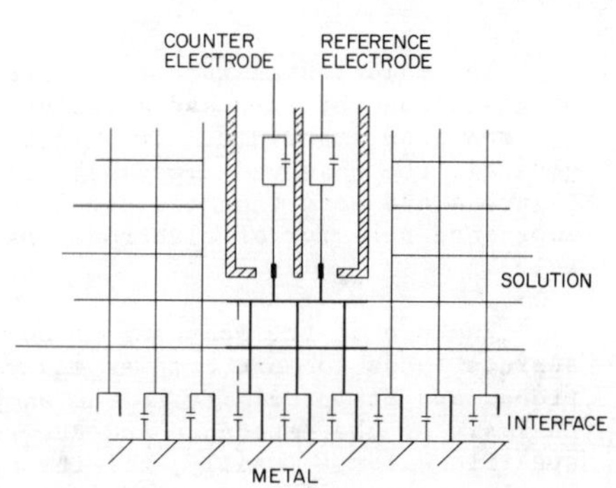

Figure 2. Equivalent circuit of the scanning probe and metal surface. The thickness of the lines are representative of the resistance magnitudes.

The schematic equivalent circuit of the cell is shown in Figure 2. The thickness of the lines corresponds to the resistance magnitudes. Figure 3 shows a simplified equivalent circuit representing the AC measurement. The reference electrode senses the potential across the interface and some solution resistance. No current flows through the reference electrode, which only senses the AC potential variations.

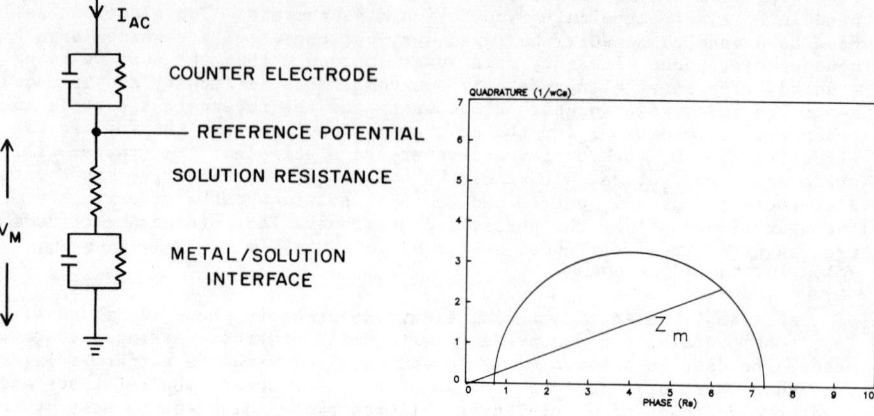

Figure 3. A simplified schematic of the circuit shown in Fig. 2.

Figure 4. An impedance diagram showing the measured impedance, Zm at a particular frequency. The impedance locus moves anti-clockwise with frequency.

The solution resistance shifts the semicircle to the right of the origin along the in-phase axis, and depending on its magnitude may dominate the measured impedance. This is often the case at high frequencies. In general, the observed impedance diagram is more complex than that shown in Figure 4 and more than one arc is observed. More detailed aspects of the impedance behavior of electrodes have been discussed in the literature (4,5).

The use of the scanning probe rather than the measurement of the entire surface leads to more complex equivalent circuits because of geometry of the probe, and its proximity to the sample being measured. This complicates the analysis of the impedance and the resolution of reactions taking place at specific sites. Ideally, all the applied current should pass into the specimen just below the orifice leading of the counter electrode. However, for a high impedance metal/solution interface, the current flows parallel to the interface with little flowing to the metal below the probe and most entering the bulk solution and then passing into the remaining freely exposed metal surface.

The potential response to an applied constant sinusoidal current (I_a) is analyzed using a lock-in amplifier. The potential response is represented by a vector (V_m) and the the lock-in amplifier separates the vector into two components, one in-phase (V_p) with the current signal, and the other, the quadrature component (V_q) at 90° out of phase. These components can also be represented by an equivalent circuit where the in-phase component is the potential across a series resistor (R_s) and the quadrature is the voltage across the capacitor (C_s). The impedance representation is given in Figure 4 by Z_m assuming the interface can be best represented by a capacitor (C_p) and resistor (R_p) in parallel, and in series with the solution resistance R_{soln}. The relations between the potentials measured by the lock-in amplifier and the equivalent series components are given by

$$R_{soln} + R_s = V_p/I_a \tag{1}$$

$$1/\omega C_s = V_q/I_a \tag{2}$$

where ω is the angular frequency related to the applied frequency (f) by

$$\omega = 2\pi f. \tag{3}$$

The values of the series components are related to the applied frequency (f) by

$$R_s = R_p/(1 + \omega^2 C_p^2 R_p^2) \tag{4}$$

$$\frac{1}{\omega C_p} = \frac{1}{\omega C_s}\left(\frac{\omega^2 C_p^2 R_p^2}{1+\omega^2 C_p^2 R_p^2}\right) \tag{5}$$

When the frequency is varied, the value of V_m changes and the locus of the impedance Z_m (= V_m/I_a) varies as shown in Figure 4. If the solution resistance is negligible the locus traces out a semicircle for a parallel equivalent circuit; where, at very low frequencies, the value of Z_m is equal to R_p, the applied current only passes across the resistor. At higher frequencies significant current pass across the capacitor, and this out-of-phase current leads to a predominant quadrature component of the equivalent circuit. At lower frequencies the interfacial in-phase component is large, while at higher frequencies the quadrature is greater than the in-phase impedance. In other words, by changing the frequency of the applied current, either the resistive or capacitative components can be emphasized.

A rough indication of the current flowing between the probe and the metal surface can be obtained from a one dimensional analysis. The equivalent circuit consists of elemental solution resistance components in series with elemental interfacial impedances between the resistances. Current is injected at the first resistance point and leaks from the end of the probe through an impedance representing the solution and surface beyond the end of the probe Z_f to ground. These problems have been dealt with in detail in transmission lines and for porous electrodes (4, 9-11). The changes in potential (v) and current (i) along the transmission line for a rectangular probe with distance (x) (allowing no current leakage from its sides) is given by

$$\frac{di}{dx} = \frac{vw}{z} \quad \text{and} \quad \frac{dv}{dx} = \frac{i\rho}{wh} \tag{6}$$

where (z) is the unit area interfacial impedance (ρ) the solution resistivity (w) the width of the probe and (h) its height above the surface. The solution to these equations give the potential, (V) and current I_a at the point of injection as

$$V = v_f \exp\left(L\sqrt{\rho/zh}\right) \tag{7}$$

$$I_a = v_f\, w\sqrt{h/z\rho}\ \exp\left(L\sqrt{\rho/zh}\right) \tag{8}$$

where v_f is the potential at the exit of the probe. The current leakage at the end of the probe (i_f), can be obtained from the above equations.

$$i_f = \frac{v_f}{z_f} = \frac{I}{z_f w}\sqrt{z\rho/h}\ \exp\left(-L\sqrt{\rho/zh}\right) \tag{9}$$

This equation demonstrates that the leakage current is dominated by the exponential term $L\sqrt{\rho/zh}$ and it can be deduced that the leakage of the current is reduced by

a) changing the size of the probe base (L);
b) decreasing the distance between the metal surface and base of the probe (h) because it increases the lateral solution resistance;
c) using a more dilute solution with a higher resistivity (ρ);
d) increasing the frequency which reduces the interface impedance.

Of these possible configurations, bringing the probe closer to the surface is the simplest to achieve and also emphasizes the necessity to have the probe as close to the metal surface as possible and to maintain a fixed distance between the probe and the surface. The other factors are dependent on the aims of the investigation and the characteristics of the metal/solution interface. For example, a specific solution may be required and a more dilute solution cannot be used, or the interfacial characteristics that are of interest can only be observed at low frequencies so high frequencies cannot be used. Increasing the probe base also has disadvantages, as it masks a larger area of the sample, making visual location of a particular site more difficult. In contrast to the difficulties, these experimental parameters can be employed to improve the resolution of defects in the metal surface.

Dividing equations 7 by 8 gives the impedance

$$Z = \frac{1}{w}\sqrt{\frac{\rho z}{h}} \tag{10}$$

when measured where the current is injected into the solution. As pointed out by de Levie, the measured impedance is equal to the square root of the surface impedance and halves the phase angle (9,11). The impedance diagram shown in Figure 4 for a metal surface would therefore be reduced, and the semicircular arc changed to one quarter of a circle.

The analysis of the effects of a circular probe as opposed to the rectangular shape analyzed above confines the current to the probe to a far greater extent. The rate of change of the current and voltage with distance, from the point of current injection, is much more rapid than the logarithmic dependence indicated by equations 7 and 8. This reduces the leakage current from the edges of the probe and would also "flatten" the observed arcs for the impedance diagram still further. The advantage gained is that the spacial resolution is increased, as the perturbation of the heterogeneities of the surface then have a smaller influence at a given distance from the point of current injections as compared with calculations using a rectangular probe.

The impedance analysis and quantitative determination of the effects of specific variations of surface impedance, for the probe used, are extremely complex. Hence, at present the observations are empirical and the techniques require standardization as a non-destructive method for observing surface heterogeneities.

In contrast to this disadvantage, the technique is extremely sensitive and it has been possible to resolve heterogeneities on all carefully prepared surfaces.

Experimental Methods

The electronic equipment for measuring variations in surface impedances is shown schematically in Figure 5. The probe shown in Figure 1 is mounted on a stage which rasters the probe over an area parallel to the surface of the sample in an electrolyte. Generally, the probe is positioned about 50 to 100μm from the surface and has a diameter at the base of 2.5 mm. Platinum wire electrodes, inserted into the two compartments act as pseudo-reference and counter electrodes. The size of the orifices at the base of the probe are about 100μm and along a diameter at 0.5 mm from the probe's center. Three PTFE "shrink tubing" sections, two inner and one outer, are used with FEP inserts to obtain a solid base except for the two orifices.

The scanning device consists of two stepping motors rotating screw drives for X and Y stages. The stepping motors are activated from a motor Control unit, controlled by a pulse output from a Hewlett Packard Model 2240 "Measurement and Control Processor," in conjunction with a Hewlett Packard Model 9845T desk top computer. These instruments are also used to acquire, store, and plot the impedance data.

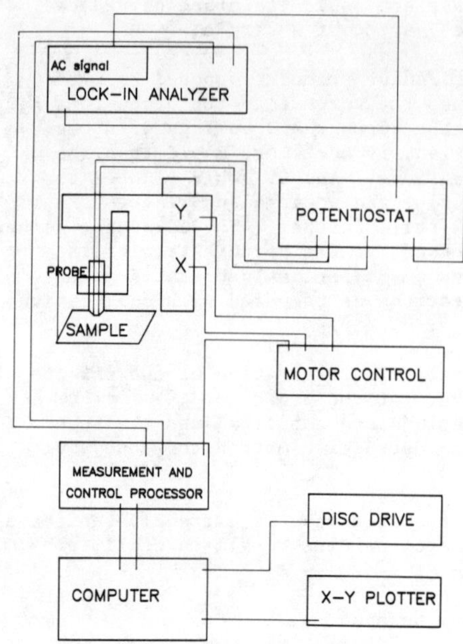

Figure 5. Instrumentation for AC scanning measurements.

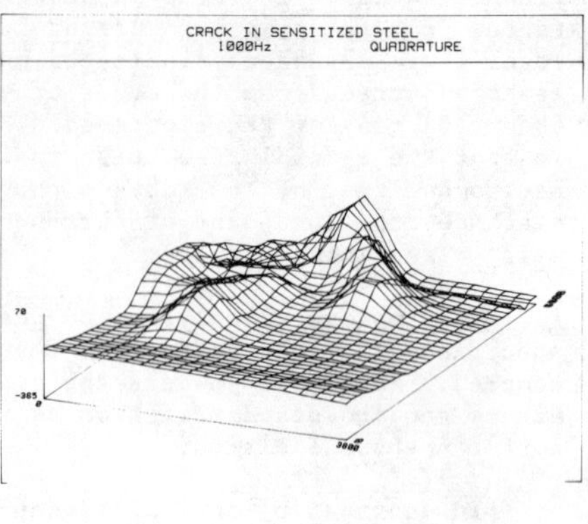

Figure 6. Variations of quadrature over a cracked Type 304 surface measured at 1000 Hz. The impedance lies between -365 = 39.48 ohm and 70 = 44.48 ohm.

The in-phase and quadrature components of the impedance are measured with a dual channel lock-in analyzer (Princeton Applied Research Model 5204) which also supplies the AC sinusoidal signal. At various times, square wave excitation is also used using a signal generator. The AC signals are converted into a current by a potentiostat (Stonehart Model BC 1200) operated in the galvanostatic mode. Generally, the voltage response is 20 to 40 mV.

Results and Discussion

Detection of Cracks on Metal Surfaces

The ability of the AC scanning technique to detect surface cracks was investigated. The Type 304 stainless steel sample was sensitized at 600°C for 24 h and intergranularly stress corrosion cracked in 6.10^{-4}M $Na_2S_2O_3$ under constant load (12). After cracking, a section of the sample was polished down to a 600 grade silicon carbide paper and then immersed in 0.1M Na_2SO_4 solution and scanned. The AC frequency was 1000 Hz. The area scanned was 3.8 by 6.8 mm.

The quadrature variations are plotted as a three dimensional surface in Figure 6 and clearly showed variations in impedance over cracks in the surface. The increase in the quadrature was indicative of a reduced surface area rather than an increased area because of the presence of the crack walls. This demonstrated the solution in the crack acted as a higher resistance path masking the walls in the crack. The greater the size of the crack, the greater the increase in quadrature. The in-phase components showed similar variations, as deduced from Figure 7 giving a contour map of the observed impedance variations. The increase in impedance was dependent on the frequency. At lower frequencies the impedances of the steel surface increased and the relative contribution of the solution resistance decreased. Under these conditions the increased area of the crack walls led to a decreased impedance.

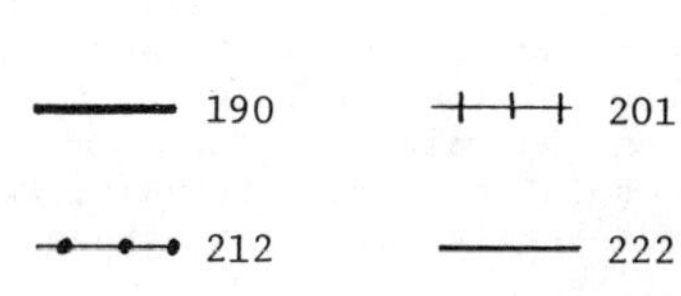

Figure 7. A contour mapping of the results shown in Fig. 6

Measurements and calculations of the impedance of porous electrodes and single pores in surfaces have been carried out (9-11). The results showed a decrease in impedance as the size of the pores increased. The work, however, differs from the scanning impedance technique as the measurements relate to the entire surface whereas with scanning only a restricted area was measured.

The results in Figure 6 show a variation of -365 to 70 units in the quadrature. These variations, about an offset value, were between 39.482 and 44.483 ohm (i.e. 3.578 and 4.0399 F respectively). These were about a 12% variation in impedance with a resolution of 0.5% on this plot.

The in-phase variations were from 19.8 to 24.4 ohm, and showed a similar resolution. In both cases, the spacial resolution of the center of the cracks was within 0.2 mm. This resolution could be improved to about 0.08 mm by decreasing the distance between measuring points during scanning. The position of the center line was clearly resolved, but difficulties in determining the dimensions of the crack opening at the surface was reduced because of the lateral spreading of the current between the base of the probe and the specimen surface. There was a correlation between the width of the cracks and the observed impedance variations as can be seen from Figure 7, which includes the contour map superimposed on a micrograph of the cracked specimen.

Welded Specimens

Measurements on welds of stainless steel, aluminum, and mild steel have been conducted. All the welds showed variations caused by the differences in the metallurgical structures because of the high fusion temperature during the welding. The aluminum and iron were tested in fluoride and sulfate solutions, respectively, which are aggressive to these metals. The impedance measurement in these studies also responded to the corrosion processes.

A Type 304 stainless steel weld was scanned in a 0.1 M Na_2SO_4 solution, and as expected, showed no significant corrosion. The weld was prepared with a Type 308 consumable insert and then filled with Type 308 weld rod using Tungsten inert-gas procedures (13). It was studied over a period of two days after the surface was wet abraded down to 600 mesh on silicon carbide paper, parallel to the length of the weld. The weld was freely exposed and not mounted on the edges coated.

The in-phase and quadrature components are shown in Figure 8 after the specimen had been in solution for about 24 h. The abrasion, it must be pointed out, did not mask the underlying structure as the weld region showed distinct impedance variations. Very few techniques are available which enable this separation to be made. The study of etched samples has not been carried out, mainly because the etching procedure would give different surface structures or roughness to which the technique would respond, complicating separation of the effects caused by etching from those resulting from electrochemical differences.

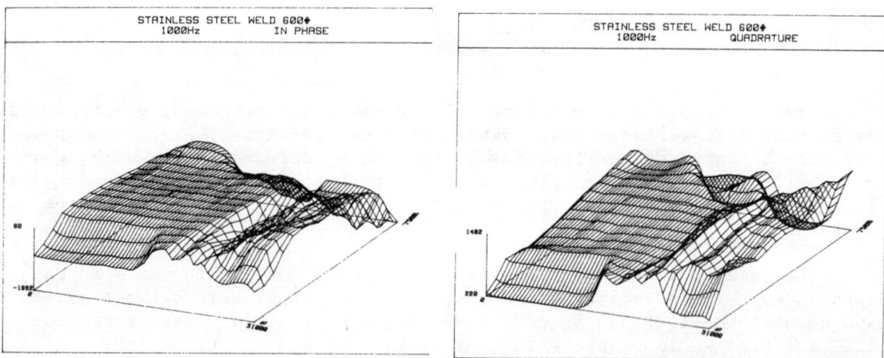

Figure 8. The in-phase (a) and quadrature (b) of an abraded stainless steel weld.

The large decreases at the top and bottom of the impedances in Figure 8 resulted when the probe was at the edge of the specimen and depended on frequency. On decreasing the frequency to 500 Hz, the in-phase impedance was higher rather than lower, as the probe moved over the edge. These major changes clearly defined the edge of the specimen, as can be seen in Fig. 9, with the contour lines for two frequencies superimposed over macrographs of the weld. The impedance changes in the vicinity of the edges also depended on the height of the probe above the surface and would also be expected to depend on solution resistivity and interfacial impedance of the sample (see equation 10). In general, the closer the probe to the surface, the greater the impedance because of the reduced area to which the current flows, and the greater the resolution of surface heterogeneities. However, the locus of the points as a function of distance above the surface for a given frequency plotted on an impedance diagram, as in Figure 4, was not linear. Above 500 Hz the in-phase component showed a maximum, as the quadrature decreased with distance from the surface.

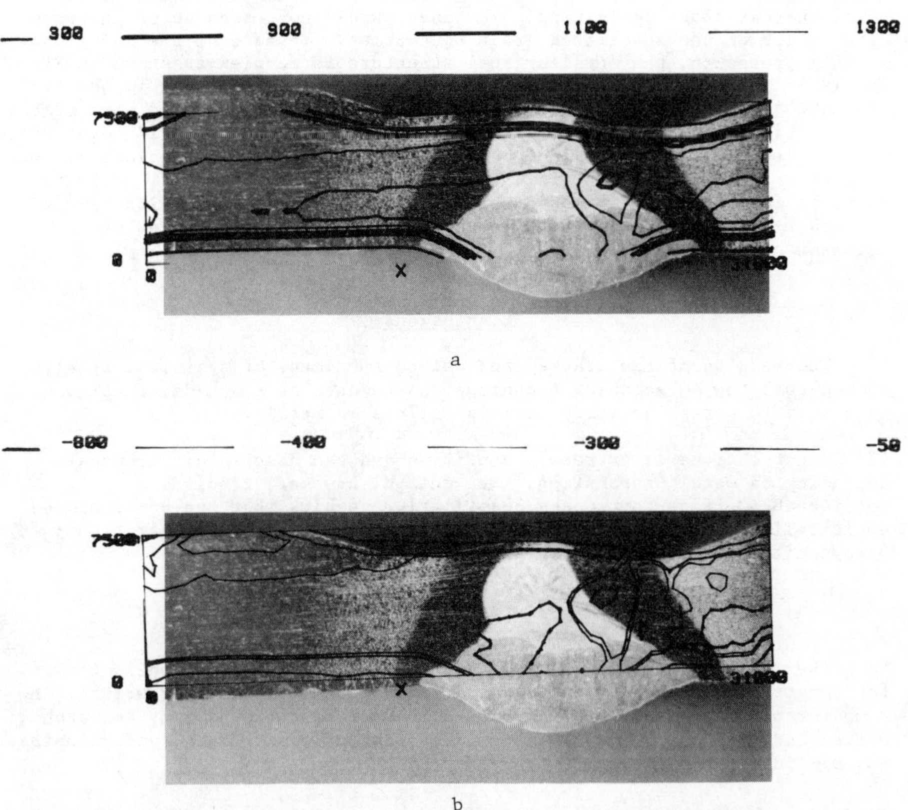

Figure 9. Macrograph of the Type 304 stainless steel weld etched after testing with contour mapping of the in-phase impedance variations superimposed (a) The quadrature variation at 300Hz and (b) the in-phase variations at 1000Hz.

The sodium sulfate solution used would not be expected to lead to corrosion as the metal becomes passive, and the rate of metal oxidation decreases with time for all areas of the surface almost independently of the differences in composition. The differences that arise because of corrosion are expected to be small, and this is brought out in the neutral sulfate solution. Nevertheless, differences can be discerned using the AC technique. The greatest differences on the weld were at the intersection of fusion lines between the bare metal, the weld metal and where the weld was molten. This can be seen on two areas in Figure 9. The second area which can be seen to give variations is the sensitized region. These do not show up very clearly in the sulfate solution, but can be defined during the early stages of exposure to the sulfate solution. If tests are required to determine the position and the extent of sensitization, more aggressive solutions should be considered. These could be sulfate solutions of lower pH, but should not require the highly aggressive solutions containing thiocyanate (14), which have been used in the DC scanning to achieve these objectives. The lower pH solutions would still be considered non-destructive, whereas those containing thiocyanate would not, because of the extensive attack of the sensitized grain boundaries. Because of the multi-pass welding procedure, the metallurgical structure is complex as shown in Figure 10. A contour map of the in-phase component is shown superimposed in the macrograph in Figure 9. The most active places on the weld were at the intersection of fusion lines between the base metal and the weld metal where the repeated melting took place on consecutive passes. These places showed very high activity when the weld was first placed in the solution after abrasion and dominated the observable impedance variations. With time of the order of 10 h, the activity decreased so that other features of the weldment could be observed.

Conclusions

(1) The results on the cracked and welded specimens of stainless steels showed that the AC scanning technique can be used as a non-destructive technique to determine structural variations in metals.

(2) The technique is extremely sensitive and can discern differences depending on metal preparation. The method, however, requires a considerable effort to produce quantitative results, but can be developed empirically for quality control e.g. to determine surface preparation homogeneity on the degree of sensitization adjacent to welds.

Acknowledgements

The author gratefully acknowledges the contribution of Donald Becker for construction of the scanning device and Yasuko Sanborn for writing the computer program. The work was funded in part by the U.S. Army Research Office and the Department of Energy, Division of Basic Energy under Contract No. DE-AC02-76CH00016.

References

1) H. S. Isaacs and B. Vyas, "Scanning Reference Electrode Techniques in Localized Corrosion," in Electrochemical Corrosion Testing , ASTM 727, F. Mansfield and U. Bertocci Edts., ASTM 1981, p. 3-33.

2) H. S. Isaacs and G. Kissel, "Surface Preparation and Pit Propagation in Stainless Steels," J. Electrochem. Soc. 119 (12) (1972) pp. 1628-1632.

3) H. S. Isaacs and M. W. Kendig, "Determination of Surface Inhomogeneities Using a Scanning Probe Impedance Technique, Corrosion, 36 (6) (1980) pp. 269-274.

4) C. Gabrielli, Identification of Electrochemical Processes by Frequency Response Analysis, Pub. Solartron Instrumentation Group, Farnborough, England, 1980.

5) D. D. MacDonald, Transient Techniques in Electrochemistry, Pub. Plenum Press, New York, 1977.

6) L. Young, Anodic Oxide Films, Academic Press, London, 1961, p. 181.

7) I. Epelboin, C. Gabrielli, M. Keddam, and H. Takenouti, "Alternating-Current Impedance Measurements Applied to Corrosion Studies and Corrosion Rate Determination," in Electrochemical Corrosion Testing , ASTM 727, F. Mansfield and U. Bertocci, Edts., ASTM 1981, p. 150-166.

8) S. Harayama and T. Tsura, "A Corrosion Monitor Based on Impedance Method" in ibid.

9) R. de Levie, "On Porous Electrodes in Electrolyte Solutions-IV," Electrochimica Acta 9 (1964) pp. 1231-1245.

10) H. Keiser, K. D. Beccu and M. A. Gutjahr, "Abschatung de Porenstruktur Poroser Elektroden aus Impedanzmessugen," Electrochimica Acta, 21 (1976) pp. 539-543.

11) R. de Levie, "Electrochemical Response of Porous and Rough Electrodes," in Advances in Electrochemistry and Electrochemical Engineering , P. Delahay Edt., Interscience Publishers, New York, 1967, p. 329.

12) R. C. Newman, K. Sieradzki and H. S. Isaacs, "Stress Corrosion Cracking of Sensitized Type 304 Stainless Steels in Thiosulfate Solutions," Met. Trans. , accepted for publication.

13) C. Czajkowski, "Intergranular Stress Corrosion Cracking of Type 304 Stainless Steel in the Spent Fuel Pool Piping of Three Mile Island - Unit 1," Report BNL-NUREG 28879, Oct., 1980.

14) B. Vyas and H. S. Isaacs, "Detecting Susceptibility to Intergranular Corrosion of Stainless Steel Weld Heat Affected Zones," in Intergranular Corrosion of Stainless Steels , ASTM STP 656, R. F. Steigerwald Ed. ASTM, 1978, pp. 133-145.

SCANNING LASER ACOUSTIC MICROSCOPY:

A New Tool For Nondestructive Testing

Carol L. Vorres, Donald E. Yuhas and Lawrence W. Kessler

Sonoscan, Inc.
530 E. Green Street
Bensenville, Illinois 60106

Ultrasonic testing has been in widespread use as a means of nondestructive testing for many years. For the most part large structures were examined for macroscopic defects due to the limited frequency range (1-10 MHz).

With advancements in materials technology a need for high resolution nondestructive test equipment developed. The Scanning Laser Acoustic Microscope (SLAM) is a high frequency ultrasonic imaging device which produces magnified images of the interiors of optically opaque materials. The extended range of frequencies 10-500 MHz enables one to inspect large as well as small components. The high frequency range allows the detection of microscopic structures down to the micron size range. Along with the high frequency capabilities, the SLAM produces images in real time (30 frames/sec) which aides in rapid characterization of defects within materials.

The SLAM technique will be described with respect to a conventional C-scan system.

Results of several applications on the SLAM are shown along with some destructive physical analysis correlating the results.

Techniques

The primary differences between SLAM and other ultrasonic testing techniques such as C-scan are concerned with image resolution, acquisition speed and image quality. For example in a C-scan system the resolution is limited by either the size of the transducer element or by the size of the transducer focal region. The C-scan technique requires mechanical raster translation of the transducer to scan a component, thus it is inherently slow. In the SLAM, objects are insonified with a continuous plane wave and then the acoustic energy is detected with a very small receiver, about 25 microns in size (1 mil). The receiver on the SLAM is a scanning laser beam, which has several desirable characteristics, for example: it can be focused to a small spot size, thus yielding a high resolution microphone. Furthermore, it can be moved very rapidly by electronic means to cover 525 lines of sample area in 1/30 second. Full grey scale pictures are displayed on a TV monitor in real time. Figure 1 is a photograph of SLAM.[1] The principles of operation have been described in the literature.[2] The instrument shown has dual operating frequency capabilities of 30 and 100 MHz. The results appearing in this paper were obtained at both frequencies, however, options from 10 to 500 MHz also exist with this type of instrumentation.

Figure 1 - Scanning Laser Acoustic Microscope (SLAM) the instrument used in these investigations. This model SONOMICROSCOPE 100 has dual operating frequencies of 30 and 100 MHz.

Located on the table top is the stage area, 100 MHz on the right and 30 MHz on the left. The 100 MHz stage resembles an optical microscope. At 30 MHz a small immersion-type tank is used for testing. Changing from one frequency to the other is accomplished by a switch and a mirror which is repositioned to redirect the laser beam probe.

These two operating frequencies (100 and 30 MHz) accommodate resolution and penetration requirements for a wide range of sample sizes. The smallest size part that can be conveniently tested at 100 MHz is approximately 40 x 40 mils (1 x 1 mm), and the largest part at 30 MHz may be many inches, depending on the material and geometric configuration. 100 MHz provides higher resolution while the 30 MHz gives greater penetration. The SLAM produces diffraction limited resolution achieving about 25 microns at 100 MHz and about three times that at 30 MHz.

Figure 2 shows two types of acoustic information provided by the Scanning Laser Acoustic Microscope. The left image is called an acoustic amplitude micrograph and the right is an acoustic interferogram. In the amplitude micrograph variations in the gray scale are of interest. What is dark is highly attenuating and the lighter areas have a high level of acoustic transmission. In the picture on the right, the straightness of the fringe lines provides information about sonic velocity variations in the material. The interferograms are also used to give a quick evaluation of flaw size since the spacing between each of the fringes at 100 MHz is 85 microns. Thus if a flaw appears on the screen, counting fringes is used to estimate size. From an imaging standpoint, the interferograms tend to enhance the edges of the flaw.

Figure 2 is presented here to illustrate image characteristics of a relatively uniform material. The micrographs are obtained at 100 MHz on a .012" thick bar of hot pressed silicon nitride. This sample contains less than 1% porosity which leads to a "clean" acoustic micrograph. Some transmission variations (diagonal striations) are apparent which are attributed to grinding marks on the surface. The fringes on the interferogram are parallel and equally spaced indicating uniform sonic velocity.

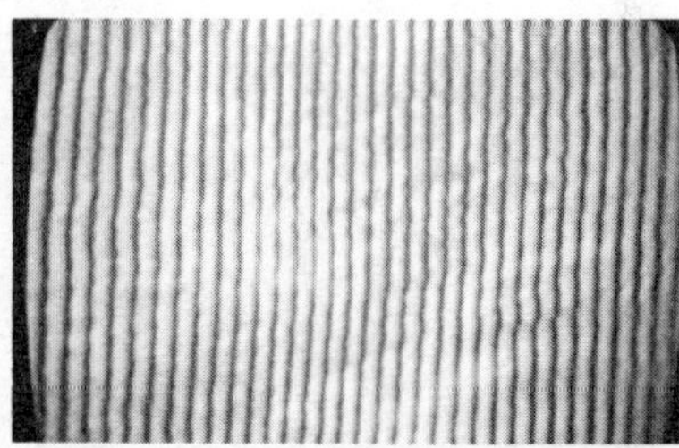

Figure 2 - Two modes of operation are shown in these 100 MHz amplitude (left) and interferogram (right) acoustic micrographs. These micrographs of hot pressed Si_3N_4 represent a relatively homogeneous material.

In addition to surface defects, the SLAM can be employed to image deep lying flaws. In this case the acoustic energy is shadowed by the defect as the (original) plane wave transmits through the sample. By making use of the various modes as well as various angles of insonification, defects cannot only be detected, but they can be characterized.

Laminar flaws and disbonded regions (parallel to the surface being inspected) produce a dark shadowed image just as a vertical (perpendicular) crack would. But by changing the angle that the incident ultrasound propagates, one can obtain depth information about these defects and also differentiate between them. The position of the shadow caused by a disbond, or laminar flaw will change with the angle of insonification, but its size will remain constant. On the other hand, a vertical crack's position will remain the same, but the shadow length will change.

Inclusions in a material will give a rise to a bull's eye pattern, due to diffraction of the ultrasound, and usually appear as a bright circular center with rings surrounding it.

Tiny cracks near the surface of a material which may have been thought to be unresolvable, based solely on the basis of the wavelength, produce an interesting phenomenon, rendering them clearly visible and therefore

WELD A

Hole Left Indicating "Good Nugget"

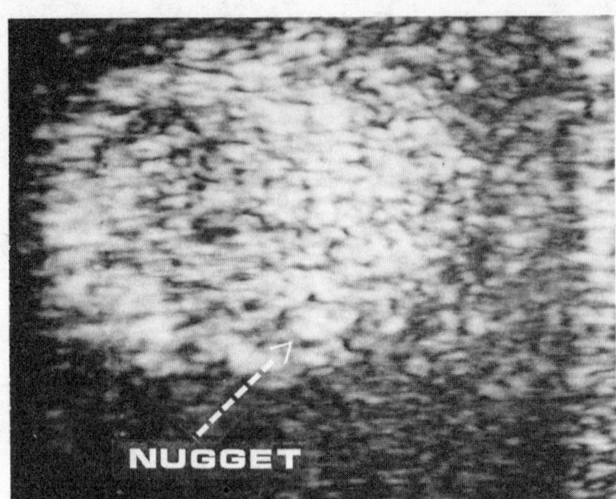

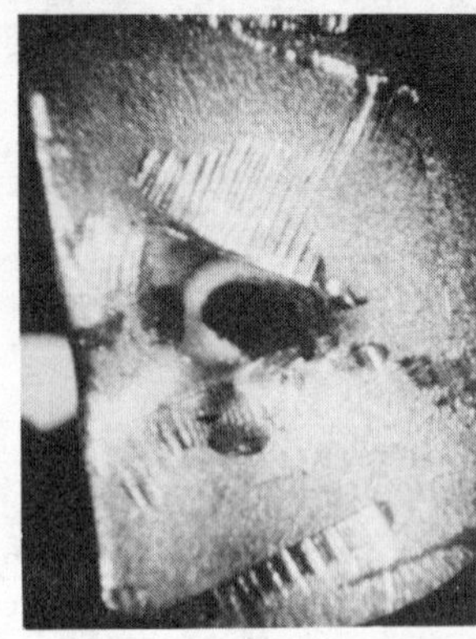

Acoustic Images

After Pull Test

WELD B

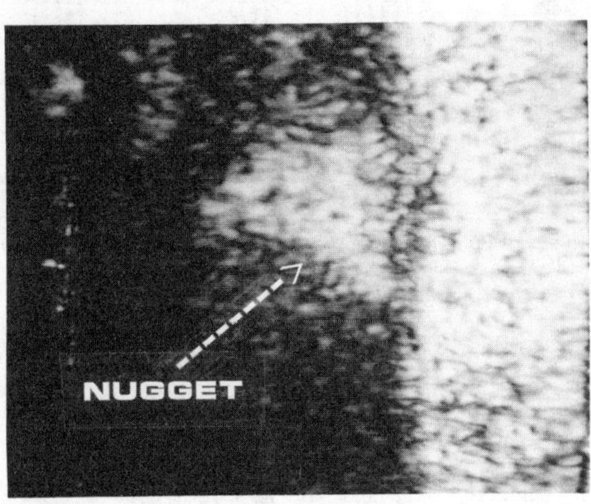

Figure 3 - Spot welds of two stamped sheet steel parts are shown here as imaged on the SLAM and then after pull testing. Both welds appeared similar upon visual inspection, but are quite different as seen on the SLAM. Weld A has a very large well-defined nugget while Weld B is rather small and irregular in shape. Pull testing confirms Weld A as a good weld by the tearing of the metal sheet. Weld B, the projection tip was not welded at all and with light force the pieces popped apart.

Acoustic Image

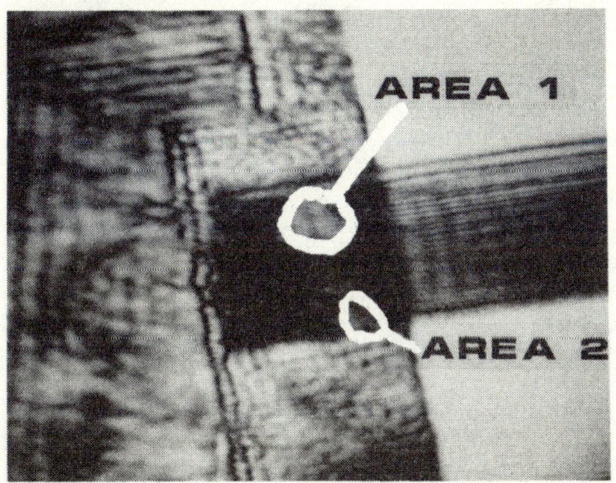

After Pull Test

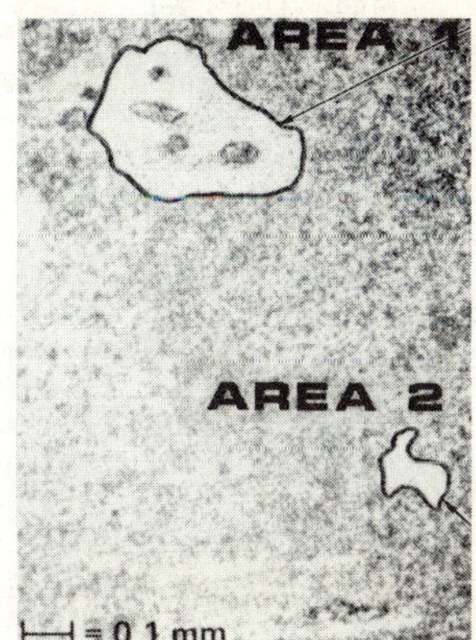

Figure 4 - A 2 mil thick silicon solar cell with a tab welded to it was examined on the SLAM at 100 MHz. The bright areas 1 and 2 correspond to a material remaining after pull test indicating bond zones.

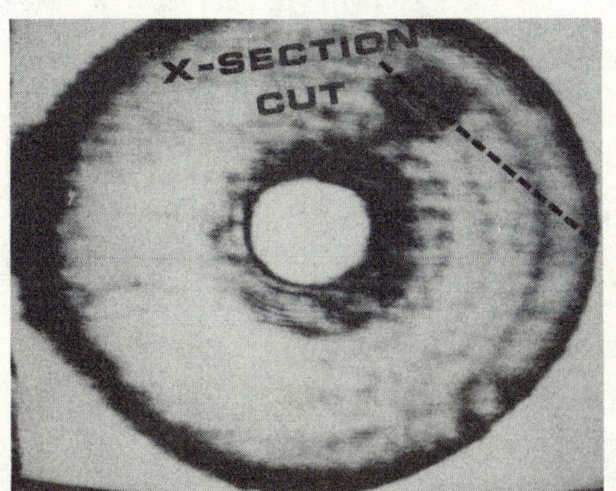

Figure 5 - A discoidal ceramic capacitor approximately 1 cm in diameter was examined on the SLAM at 30 MHz and found to contain a delamination. Cross sectioning of the capacitor revealed the delamination in the area indicated.

detectable. In particular, the acoustic wave traveling through the material hits the small crack and is partially converted into a surface wave. The original sound wave then interferes with the surface wave producing rings which seem to eminate in a cone-like pattern from the flaw. Following the pattern back to the apex of the cone is found the tiny crack itself.

Most anomalies fall into one of the above mentioned categories, namely, cracks, delaminations, or disbonds, voids, inclusions or surface pits. Users of acoustic microscopes have done extensive correlative analysis with the acoustic images to verify their results. Those shown in Figures 3 and 4 involve the use of SLAM to evaluate spot welds in metal parts going into a kitchen appliance, and also evaluating a weld on a solar cell for a satellite array. A common application, Figure 5, involves looking inside ceramic capacitors for physical defects that have not yet become evident electrically. Due to the laminar nature of these ceramic capacitors, they are prone to delaminations.

Verification of the acoustic images in all cases required destroying the part either by pulling apart or cutting it open. Therefore, the SLAM can be employed in the job of quality assurance or in quality control when inspection of a large number of components must be screened.

In conclusion, various applications of the SLAM have illustrated its versatility in nondestructive quality control testing. The defects in materials cannot only be detected but also characterized, which would provide an overall better understanding of the strengths or weaknesses of a material or component. The images produced are high resolution and produced extremely rapidly (30 frames per second), lending this technique for use not only in the analytical lab, but also for automated testing.

References

1. Commercially available under trade name SONOMICROSCOPE™ 100, Bensenville, IL 60106

2. Kessler, L.W. and Yuhas, D.E., "Acoustic Microscopy 1979", Proc. IEEE, 67, (4) pp. 526-536 (1979)

PHOTOACOUSTIC MICROSCOPY FOR NDE OF OPAQUE SOLIDS

R. L. Thomas, L. J. Inglehart, L. D. Favro and P. K. Kuo

Department of Physics
Wayne State University
Detroit, MI 48202

The technique of scanning laser photoacoustic microscopy (SPAM) using a gas-filled cell is described. It is shown to be a useful tool for the nondestructive evaluation (NDE) of surface- and near-subsurface-flaws in opaque solids. In the case of surface flaws, detection is primarily the result of spatial variations in optical reflectivity, and resolutions of better than 6 μm are demonstrated. For subsurface flaws, detection is the result of thermal wave scattering, with probe depth and lateral resolution controlled by the thermal diffusion length in the solid. The importance of the phase of the photoacoustic signal is emphasized and the phase signatures of open as well as closed vertical and lateral subsurface cracks are described theoretically and compared with experimental data.

Experimental Technique

A block diagram of the experimental arrangement is given in Fig. 1. An Ar-ion laser is used as the thermal wave source. Its intensity is square-wave-modulated at an audio frequency and focused on the surface of the sample to be scanned. The sample is contained in a sealed, gas-filled acoustical cell, the pressure of which is monitored by means of a miniature microphone. The position of the focal spot is scanned over the surface of the sample by means of microprocessor-controlled stepping motor stages. The ac component of the cell pressure is preamplified and phase-sensitively detected by means of a lock-in amplifier. The output of the lock-in amplifier consists of two dc voltage levels, one of which is proportional to the magnitude, the other to the phase of the ac signal, and is stored in the memory of the microprocessor after analog-to-digital conversion (see Fig. 2). The step size of the x- and y-stepping stages is in multiples of 6.35 μm, and a complete SPAM scan consists of typically 10,000-20,000 data points. This block of data is transmitted to a microcomputer for access to standard contour and line-perspective drawings. The memory of the microprocessor can also be read rapidly through a digital-to-analog converter, the output of which modulates the intensity of a cathode ray tube (CRT) to obtain a gray-scale photoacoustic micrograph.

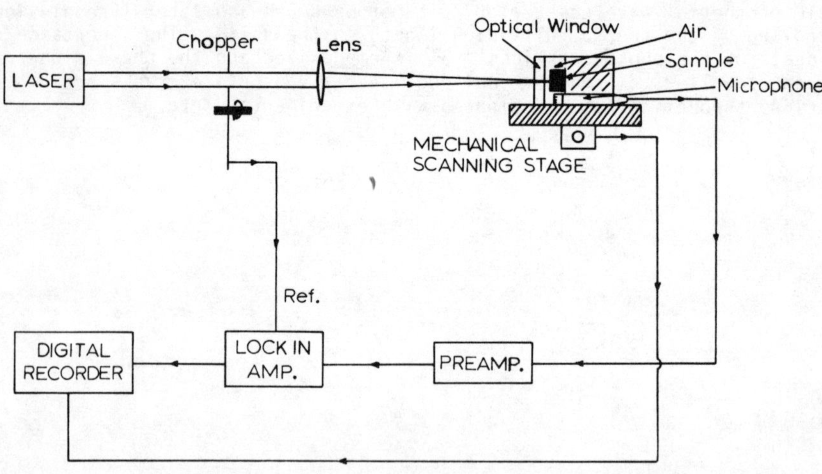

Figure 1 - Schematic diagram of the photoacoustic scanning system.

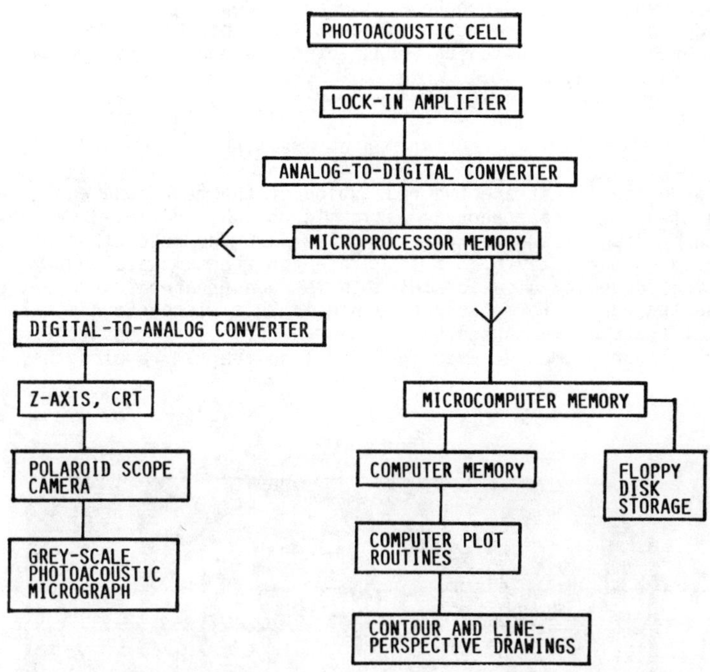

Figure 2 - Flow chart representing data acquisition and display.

Analytical Description

A detailed description of the theory of photoacoustic imaging using gas-filled cells is given elsewhere (1-3). Essentially, the intensity-modulated focal spot of the laser acts as a time varying thermal source. The solution to the one dimensional thermal diffusion equation which is appropriate for a time-varying source may be written as

$$T = T_0 \exp[i(qx - \omega t)], \qquad (1)$$

where $q = (1 + i)/(\omega\rho c/k)^{1/2}$, $f = \omega/2\pi$ is the modulation frequency, ρ is the mass density, c is the specific heat capacity, k is the thermal conductivity of the sample, and T is the sample temperature. This wave solution, with equal real and imaginary parts for its wave number, is very highly damped, being attenuated by a factor of about 500 in a depth of one thermal wave length beneath the sample surface. It is just this characteristic which is responsible for the usefulness of SPAM as an NDE probe of the near-subsurface region of the sample. By varying the modulation frequency over the audio range of the microphone, the probe depth of the thermal waves can be varied up to a maximum depth of about 1 mm using the magnitude of the signal and about 2 mm using the phase of the signal for a typical material, say for example, aluminum.

Experimental Examples

In order to illustrate the resolution of the technique and give an example of a gray scale photoacoustic micrograph, in Fig. 3 we show a SPAM micrograph of an integrated circuit chip (4). The resolution of this micrograph is one step (6.35 μm). Since the thermal wavelength at the modulation frequency used for this scan is much greater than 6 μm, one might conclude that the information contained in this micrograph is strictly surface in nature and caused by changes in optical reflectively. However, for this planar geometry, even for very long thermal wavelengths, thermal

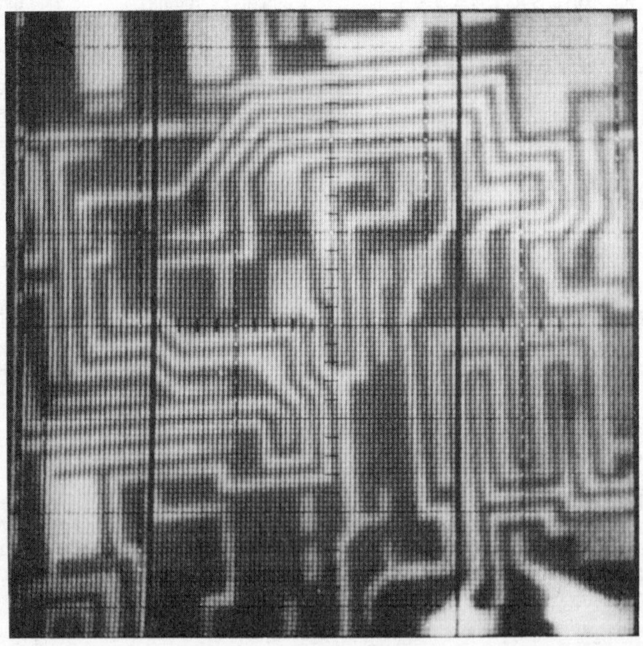

Figure 3 - SPAM micrograph of an integrated circuit chip (4). The field of view is 813 μm x 813 μm, with a resolution of 6 μm.

reflectively variations from subsurface boundaries can often dominate and the lateral resolution of the thermal wave images is determined not by the thermal wave length, but by the depth of the interface causing the variation in thermal reflectivity. Consequently, subsurface delaminations of planar structures such as this example are particularly amenable to NDE by SPAM. Furthermore, a careful study of the phase variation can provide information regarding the type of discontinuity in thermal properties at the interface and its depth beneath the surface. For instance, a void gives a very different characteristic phase signature than does a closed crack.

In Figs. 4 and 5 we show an example of photoacoustic detection of a back surface slot (analogous to a void in a continuous sample). The aluminum sample was fabricated with a back surface slot which tapered from thermally thin at one end (∿25 μm) to thermally thick at the other end (∿1 mm). Line scans of the photoacoustic magnitude and phase are shown in Fig. 5. A detailed description of this experiment and the associated three dimensional thermal diffusion theory is given elsewhere (1). Good quantitative agreement is found between theory and experiment, including the 45° phase lag of the photoacoustic signal in the thin region relative to that in the thick region.

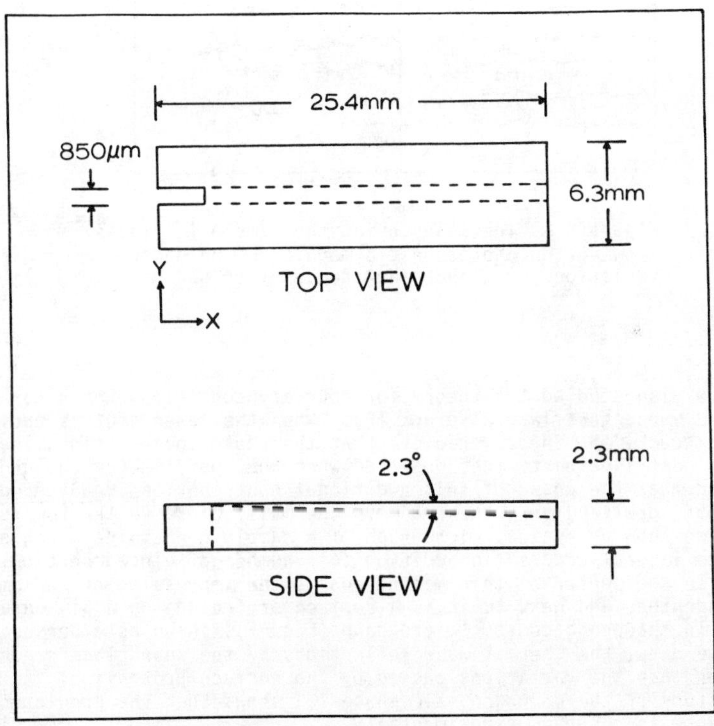

Figure 4 - Aluminum slab with tapered back-surface slot (1).

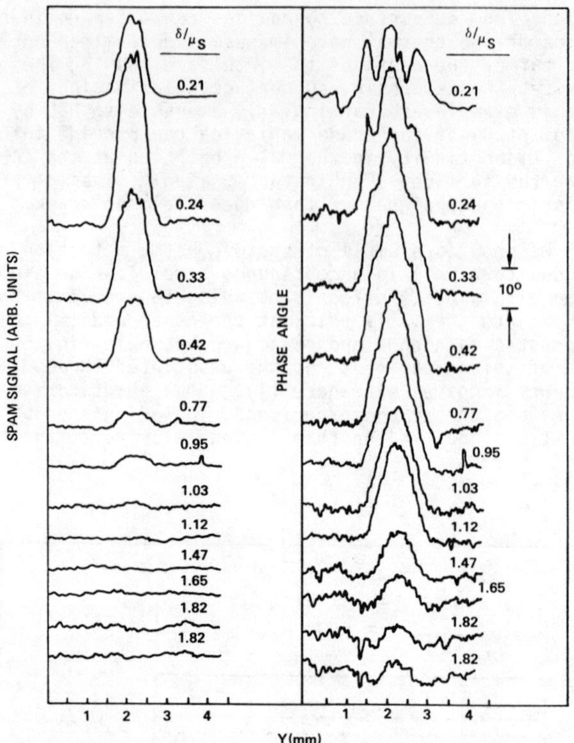

Figure 5 - Line scans using the sample in Fig. 4 showing the photoacoustic magnitude and phase variation for a chopping frequency of 600 Hz

We have also studied the theory for boundary conditions which correspond to a closed subsurface lateral crack (2). When the laser spot is above the subsurface crack, the theory predicts that the photoacoustic signal has an additional contribution to that observed when the spot is over the undamaged region, and that the phase of this additional contribution is <u>advanced</u> in phase by 45°, provided that the crack is thermally close to the top surface. In Fig. 6 we show an optical micrograph of a sample containing a region of such closed lateral cracks (in addition to a number of closed vertical cracks). In the center of this micrograph is the impression of a diamond (Knoop) indentor, and next to it is a surface protrusion on a SiC ceramic surface. In the photoacoustic micrograph (Fig. 7), shown as a perspective plot of the data, the thermal wave reflections by the subsurface cracks are much larger than the variations caused by the surface protrusion. A detailed study of the photoacoustic phase (2) shows that the predicted 45° phase advance is observed experimentally.

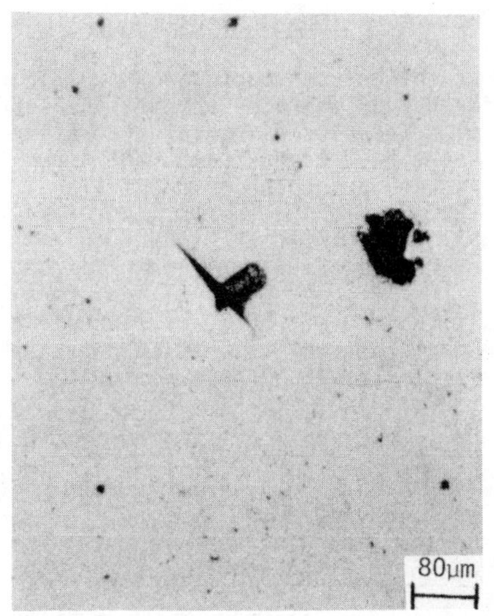

Figure 6 - Optical micrograph of a diamond Knoop indentation (center) with a large side chip, and a surface protrusion (right) on a SiC ceramic surface.

Figure 7 - Computer generated perspective plot of 1 kHz photoacoustic data from a .635 x .635 μm region centered on the above Knoop indentation (Fig. 6), using a 6.35 x 6.35 μm pixel.

Conclusions

The technique of scanning laser photoacoustic microscopy (SPAM) is shown to be a useful tool for NDE of surface- and near-subsurface-flaws in opaque solids. Subsurface voids and closed lateral cracks have very different phase signatures which can be used to characterize the flaws and obtain information regarding their depth beneath the surface.

Acknowledgments

The authors thank K. Grice, J. Lhota, K. T. Chen, Z. J. Feng, and I. Bergel for assistance with the data collection and analysis. This work was supported by ARO under Contract No. DAAG29-81-K-0113.

References

1. R. L. Thomas, J. J. Pouch, Y. H. Wong, L. D. Favro, P. K. Kuo and Allan Rosencwaig, "Subsurface Flaw Detection in Metals by Photoacoustic Microscopy," Journal of Applied Physics 51 (2) (1980), pp. 1152-1156.

2. P. K. Kuo, L. D. Favro, L. J. Inglehart and R. L. Thomas, "Photoacoustic Phase Signatures of Closed Cracks," Journal of Applied Physics 53 (2) (1982), pp. 1258-1260.

3. P. K. Kuo and L. D. Favro, "A Simplified Approach to Computations of Photoacoustic Signals in Gas Filled Cells, Applied Physics Letters (in press).

4. L. D. Favro, P. K. Kuo, J. J. Pouch and R. L. Thomas, "Photoacoustic Microscopy of an Integrated Circuit," Applied Physics Letters 36 (12) (1980), pp. 953-945.

A MEASUREMENT OF CRACK DEPTH BY CHANGES IN THE FREQUENCY SPECTRUM OF A RAYLEIGH WAVE

A. J. Testa and C. P. Burger

Department of Engineering Science and Mechanics
and Engineering Research Institute
Iowa State University
Ames, IA 50011
USA

The property of a broadband Rayleigh surface wave, according to which its frequency spectrum varies with depth below the surface, is used to measure the depth of rough open shallow fatigue cracks in steel. If the crack depth is less than 1.5 wavelengths (1.5 λ) of the lowest frequency components of the incident wave, its depth, normal to the surface, can be measured from 1 mm to 1.5 λ with a resolution of 0.2 mm. It is shown that the open portion of a crack acts as a filter which passes frequency components with wavelength longer than 0.8 x crack depth. The measurement is independent of the roughness of the crack, penetrant or oil inclusion in the crack, weld splatter, oil or dirt on the surface and residual surface or crack tip stresses. If the crack contains a closed portion beyond the region already measured, the additional length, along the crack, can be measured from the same ultrasonic signals by time of flight measurement between any two of three clearly identifiable R-waves in the complete time scan.

Introduction

Previous studies to determine the depth of open shallow slots and slits considered the effect that such "defects" have on the frequency spectra of the transmitted or "forward scattered waves" [1-4]. It was shown that when the slots are shallower than the depth of the lower frequency components in an incident surface wave, the wave is divided into two parts. The deeper portion of the incident wave diffracts at the tip of the slot or slit and appears, ultimately, as a low frequency Rayleigh wave leading the shallower portion of the incident Rayleigh wave that does travel around the defect.

Since surface breaking fatigue cracks represent a large body, if not the largest body, of structurally dangerous defects, it was decided to investigate how well the earlier procedures, as described above, could work in determining the depth of such cracks. The work reported here is specifically concerned with measuring the depth of "open" cracks with depths less than the depth of the incident Rayleigh wave.

Many techniques for interrogating surface cracks have been developed through the years. Most of them are based on time of flight measurement of acoustic waves with some others using amplitude measurements and very few using frequency information [5-12]. These techniques have used various wave forms for their purposes; however, it is believed by the writers that surface waves are better suited for defects close to the surface. Therefore, they have been selected for the research herein presented.

The principal aim of this investigation is to establish an inspection method which can reliably size surface cracks and, in addition, remain relatively insensitive to most operational variables that affect the existing methods [13-18]. Some supporting work has been done in the past which opened the doors to the method described in this text [1,19-22]. More recent work has been presented at various conferences [2-4]. This paper presents recent progress in the characterization of open cracks.

The Rayleigh Wave

These highly energetic waves have been selected to do all of the continuing research because they have some outstanding properties. Surface waves contain most of their energy confined to a layer about 1 to 1-1/2 wavelengths in depth from the free surface. In addition, they exhibit, in the case of a broadband Rayleigh wave, a range of frequencies at the surface. Figure 1a represents a typical time scan of a Rayleigh wave and Fig. 1b is the corresponding frequency spectrum at the surface. This spectrum will lose its higher frequencies as the wave goes deeper into the material because the higher frequencies cannot penetrate as deeply as the lower frequencies.

Another good characteristic of these waves is their capability to travel long distances with relatively small attenuation. They are, also, nondispersive as long as they travel on a flat surface.

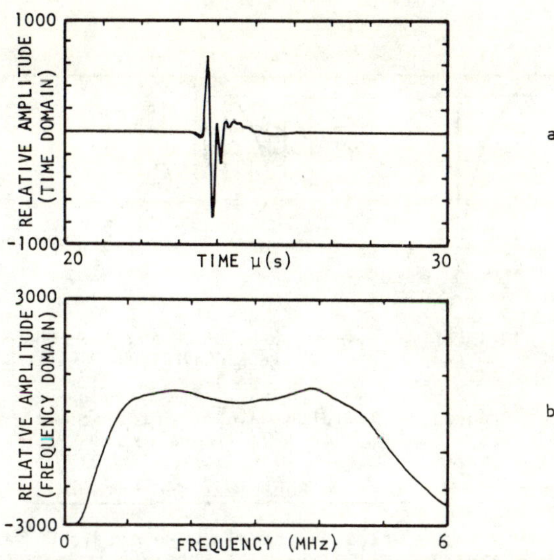

Figure 1. a. Reference Rayleigh wave
b. Reference frequency spectrum

Rayleigh wave interaction with a slot

The displacements of a Rayleigh wave, as predicted by theory [23-25], decay exponentially with depth. Because of the proportionality that exists between displacement and energy in elastic systems, it is considered a good assumption to represent the energy decay of such a wave in an exponential form. Figure 2 depicts the interaction process of a Rayleigh wave with a slot using the suggested configuration for its energy distribution with depth. Stage I is representative of the energy distribution of a typical incident Rayleigh wave. This wave is called for short the incident AFR, which stands for All Frequency Rayleigh. Stage II shows very specifically the generated Rayleigh waves upon interaction of the incident AFR. Much energy is lost to mode conversions to P- and S-waves, but this is not depicted in the figure. The AFR energy that impinges on the slot generates a reflected All Frequency Rayleigh wave, called AFR_r, and a transmitted All Frequency Rayleigh, AFR_t. The latter travels along the slot face and, eventually, around the slot and back up to the surface. However, the lower energy of the original AFR, which did not directly impinge on the slot, mode converts to some other wave which eventually reaches the surface and reconverts back into a Rayleigh wave. This wave which generates from the lower frequency deep energy in the incident AFR is called LFR_t, or Low Frequency Rayleigh, because its spectrum, as will be shown later, does not contain the higher frequencies present in the original wave at the surface. Hence, the total time scan of the transmitted package of signals, Stage III, should contain two distinct waves, the LFR_t and the AFR_t. These two waves are usually separable, and experiments have been conducted to check the validity of the interaction scheme described for Fig. 2 with very encouraging results [26].

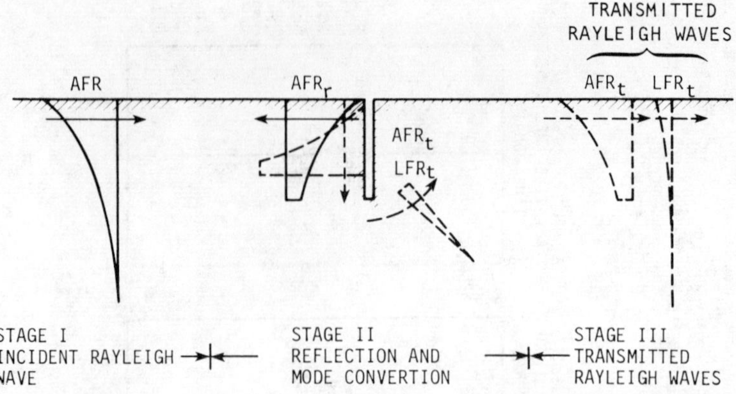

Figure 2. Rayleigh wave interaction with a slot

Testing Equipment and Materials

The results to be presented herein were obtained from an actual fatigue crack grown in A387-74A-Gr.22 steel with 2-1/4% Chromium and 1% Molibdenum. The testing probes were nominal 5.0 MHz P-wave transducers broadbanded between 0.5 - 5.5 MHz as shown in Fig. 1b. The spectrum of the waves generated can be modified by means of electronic damping provided in the signal generator. For this reason a central frequency is not obvious in Fig. 1b. Rayleigh waves are generated by mounting the transducers on a 90° Lucite wedge.

All the tests are run by placing the transmitter and the receiver in line at equal distances on either side of the crack (in this case, 15 mm on each side). The signal processing is accomplished by means of an LSI-11 microprocessor with built-in programs to handle the analysis of the data.

Experimentation

It is important to observe whether or not the behavior of an open crack is the same as that of a slot (previously discussed). In addition, it is necessary to seek information on the closed portion of a partially open fatigue crack.

Figures 3a and 3b depict the fatigue crack that was tested. After it was grown down to point b Fig. 3b, the crack was opened by applying an external load that caused sufficient permanent yielding at the crack tip to keep the crack fully open after removal of the load. After this, the specimen was further fatigued to extend the crack from point b to point a in Fig. 3b. Upon removal of the cyclic load, the portion from a to b was tightly closed (invisible to the eye). The gap in the closed portion of the crack was estimated to be less than 5μm.

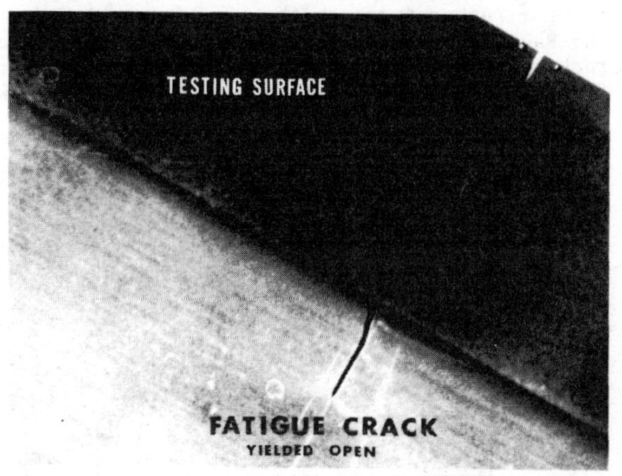

Figure 3a. Fatigue crack in A387 steel

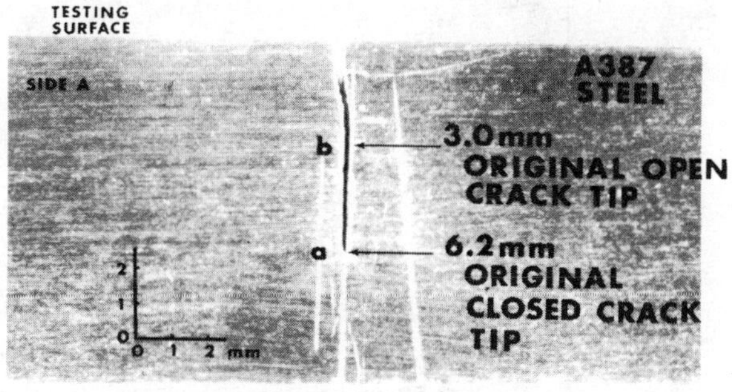

Figure 3b. Side view of fatigue crack in A387 steel

This partially closed crack was first tested without cleaning the specimen; i.e., the crack was polluted with cutting oil and steel particles. Then the sequence of signals shown in Fig. 4 were obtained. Figure 4a is a sampling of all the transmitted waves. It includes the Rayleigh wave which derives from the shear that is reflected from the far surface below the crack as described in reference 3. The group of waves that has been studied for the slot and slit cases is shown on the Transmitted Wave Group. An expansion of these waves is shown as the solid line in Fig. 4b. Two new waves can be identified that were not present with slots or slits. These are the Thru Transmitted Rayleigh (TTR) and the All Frequency Rayleigh generated at the tip of the closed portion of the crack; i.e., point a, called the (AFR_a). The TTR represents the portion of the energy from the incident Rayleigh wave that passed through the crack using as coupling medium the impurities contained in the crack. Wave AFR_a is that portion of the run around the crack Rayleigh wave that does not turn at the tip of the open crack (point b), but rather turns at the tip of the closed crack (point a). Wave AFR_b is the Rayleigh wave that turns around at the tip of the open portion of the crack. A preliminary check on these statements can be performed by using the time of travel technique to measure paths followed by the waves.

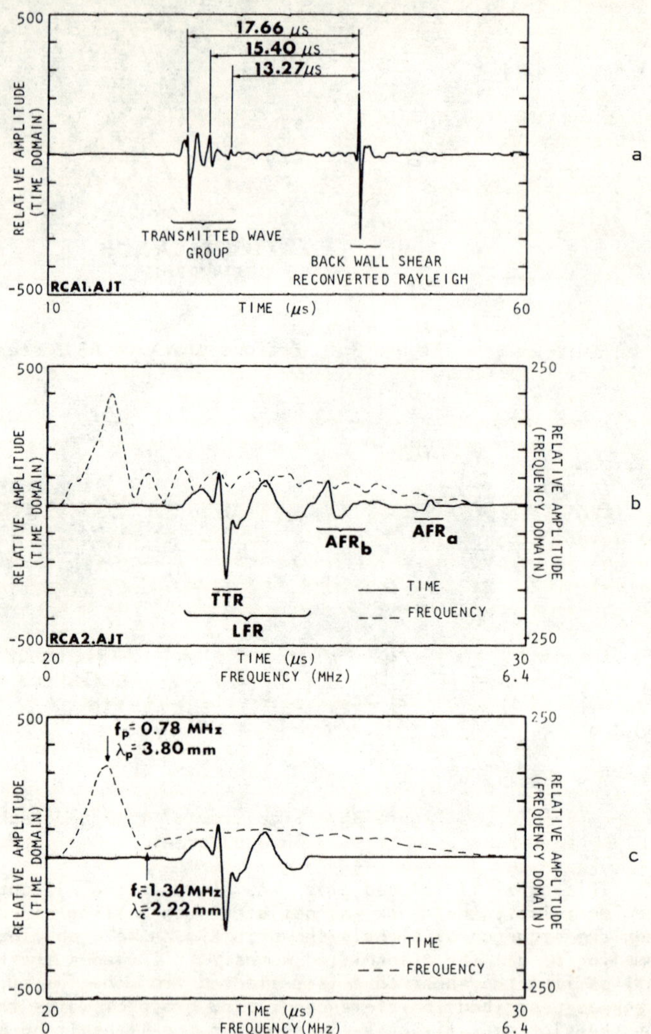

Figure 4. Sequence of samples of polluted fatigue crack under no load
a. Full view of transmitted signals
b. Transmitted wave group and FFT
c. Leading wave group and FFT

The time difference between the TTR and the AFR_b (Fig. 4a) should be twice the depth of the open portion of the crack. Knowing that the Rayleigh wave velocity in this material is approximately v_R = 3.0 mm/μs, one can readily establish that the path traveled was d_{open} = 1/2(17.66 - 15.4)(3.0)= 3.4 mm. This value should agree with the measured external open portion of the crack which is shown as 3.0 mm on Fig. 3b. Since the internal depth of the open portion of the crack is likely to be more than 3 mm, the time domain data confirms that the AFR_b and TTR waves, indeed, are Rayleigh waves following the described paths.

Next, the time between the AFR_b and AFR_a should represent the time needed by the AFR_a to travel twice the additional path a - b. From time measurements this path can be calculated to be d_{closed} = 1/2(15.4 - 13.27) (3.0) = 3.2 mm. This additional crack length checks well with the externally measured closed increment of 3.2 mm shown in Fig. 3b.

Among the waves observed in Fig. 4b, one can still distinguish the strong LFR, but it is now polluted by the TTR. It leads both the AFR_a and the AFR_b so presumably the LFR is derived from the "cut-off" at the open tip b. As was mentioned earlier, the LFR diffracts at the crack tip and manifests itself as a Rayleigh wave on the surface upon reconvertion. Since it actually leads the TTR, which, as a through-transmitted wave, travels the shortest of all the possible paths, it could not have propagated as a Rayleigh wave all the way. A mode conversion must have occurred at the tip which then reconverted to a Rayleigh wave at the top surface, some distance away from the crack itself. The question whether it is, indeed, a Rayleigh wave is answered later in this paper.

The solid line in Fig. 4c represents the LFR combined with the TTR. The dashed line is the frequency spectrum of these waves only. From this type of spectrum, the peak frequencies (f_p) and the cut-off frequencies (f_c) are selected). If the frequency spectrums of the waves in Fig. 4c and 4b are compared, one finds that the spectrum of the LFR is very dominant.

The next step was to load the specimen in bending and check the variations in the signals to establish with more certainty their origin. All the figures from here on are arranged in the same manner as Figs. 4.

Upon the application of a 1000 lb load at the center span of the specimen (opposite to the crack), the scans in Fig. 5 were obtained. There is no obvious shift of signals in Fig. 5a when compared with Fig. 4a. Figure 5b does indicate a slight reduction in the amplitude of the TTR; and Fig. 5c indicates a shift of the cur-off, f_c, frequency toward a lower value which indicates a deepening of the diffraction point.

In references 3 and 4 curves were developed for predicting the depth of slits and slots from a knowledge of the cut-off frequency (f_c), or more precisely, the cut-off wavelength ($\lambda_c = v_R/f_c$). Later this sequence of tests will be used in the same manner, and the two curves will be compared. If the values obtained so far for the cut-off wavelengths are used with the slot prediction curves, Fig. 11, one will find that the curves yield values in the order of 3.0 to 3.2 mm. This indicates that the LFR was generated at the open tip and not at the closed tip of the crack. Therefore, this shows that the method may be insensitive to closed cracks, which is not the most desirable fact. However, one must recognize two things at this stage: first, the time of travel between the two tips can still be used; and second, the method is being tested at its limits (a +6.0 mm deep crack).

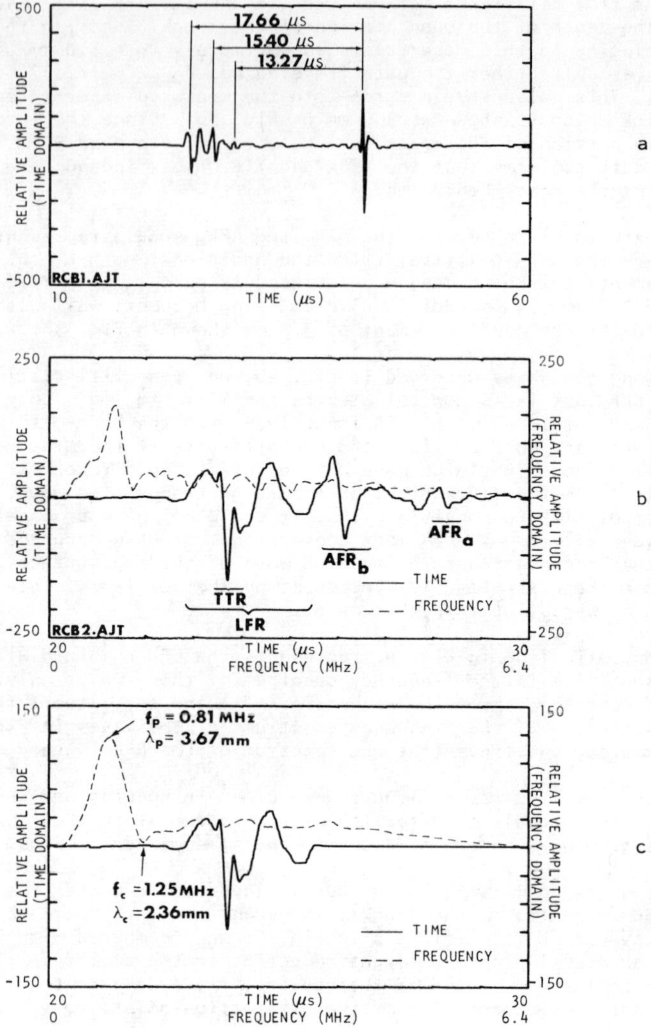

Figure 5. Sequence of samples of polluted fatigue crack under a 1000 lb flexural load
a. Full view of transmitted signals
b. Transmitted wave group and FFT
c. Leading wave group and FFT

The signals depicted in Fig. 6 were obtained upon increasing the load to 2000 lbs. From Fig. 6a it is observed that the AFR_a has become the dominant run-around-the-crack Rayleigh, while the AFR_b has almost disappeared. In addition, the LFR has broadened more which is an indication of it being generated at deeper levels. The TTR is very weak which indicates that the coupling bond between the crack walls is not so good any more. Figure 6c shows what is believed to be the LFR and its frequency spectrum. Obviously, it is very difficult to choose any particular frequency due to the many modulations which are present. This may be caused by the fact that at this moment there may be a mixed diffraction created by the polluting material and the actual crack tip at those lower levels of the crack, therefore creating additional phase cancellation problems.

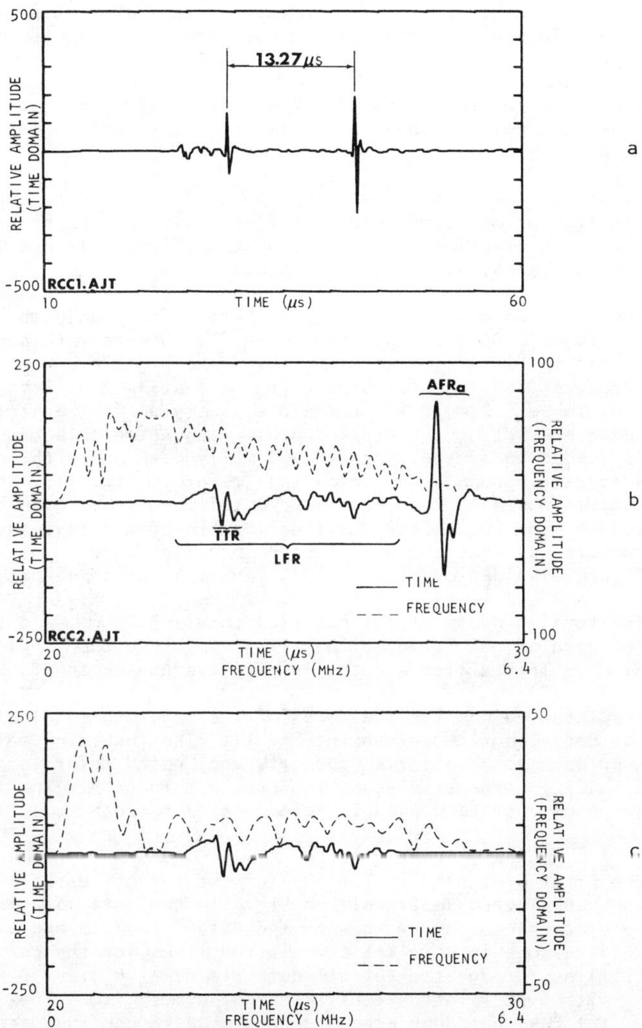

Figure 6. Sequence of samples of polluted fatigue crack under a 2000 lb flexural load

a. Full view of transmitted signals
b. Transmitted wave group and FFT
c. Leading wave group and FFT

Figure 7 represents the time scan of the signals when the specimen is loaded to 4000 lb. It is observed that the TTR has disappeared completely, and the portion containing the LFR has spread over a longer time interval (Fig. 7b). At this point the crack was open to about 1 mm wide at the top, and its length to the tip "a" could easily be tracked visually. Permanent yielding at the tip, however, was small if any at all. The frequency spectrum of the LFR shown in Fig. 7c yields a cut-off wavelength λ_c = 4.34 mm which if used in conjunction with the equations for slot data in Fig. 11, yeilds a crack depth of only 5.3 mm rather than the measured external value of 6.2 mm. This difference will be explained later as the rest of the data is analyzed.

The crack was then cleaned while holding the specimen under load to remove as much pollutants as possible, then the next series of tests were performed.

First the load was removed and the crack allowed to close to its original configuration (same as used for Fig. 4). Figure 8 shows the results. One's first observation is that though transmission is now small, the TTR has about vanished, as should be expected.

The specimen was then loaded, as before, to double the former load. With a load of 8000 lbs, the results in Fig. 9 were obtained. They should be compared to Fig. 7, especially Fig. 9b to Fig. 7b. When the depth is estimated from the line for slots (Fig. 11) at λ_c = 4.26 mm, the depth again comes out to be 5.2 mm. It, therefore, seems as if the slot curves may not be good in predicting the crack depths. To settle this argument the specimen was loaded until the crack stayed fully open and with enough permanent strain to remain open after unloading (refer to Fig. 3). The top surface of the specimen was then machined incrementally to leave even shorter cracks as depicted in Fig. 10. After the final machining and testing, the specimen was fractured so that the actual crack profile could be obtained (Fig. 10). This figure also depicts the various levels at which samples were taken.

The total amounts of cut material totals 3.79 mm, and the computed average depth of the failure line with respect to the final machined surface is 2.61 mm. This yields a total initial average depth of about 6.4 mm.

Predicted values for the depth of the crack were calculated by using the slot data equations given in Fig. 11. The resulting values are tabulated in Table 1. The terms "cut-off" and "peak" refer to the points f_c, λ_c and f_p, λ_p, respectively, as in Figs. 4 through 9. The terms are according to the nomenclature used in references 1 through 4, 19 through 22 and 26.

The error presented in Table 1 for the cut-off data is linear with depth. When plotted separately on Fig. 11, the data for the open (yielded) fatigue crack can be compared with the data for slots and slits in that they all yield reasonably straight correlation lines for the cut-off values. A best fit line through the cut-off data yields $\lambda_c = 1.2 + 0.48d$ or $d = -2.5 + 2.08\ \lambda_c$ with a correlation coefficient of almost 1.0. Peak values of wavelength (λ_p) have not been emphasized here, although they are plotted in Fig. 11, because it has been found that they are very sensitive to testing conditions. That is not the case for the cut-off values. The cut-off line is relatively insensitive to typical variations in test conditions.

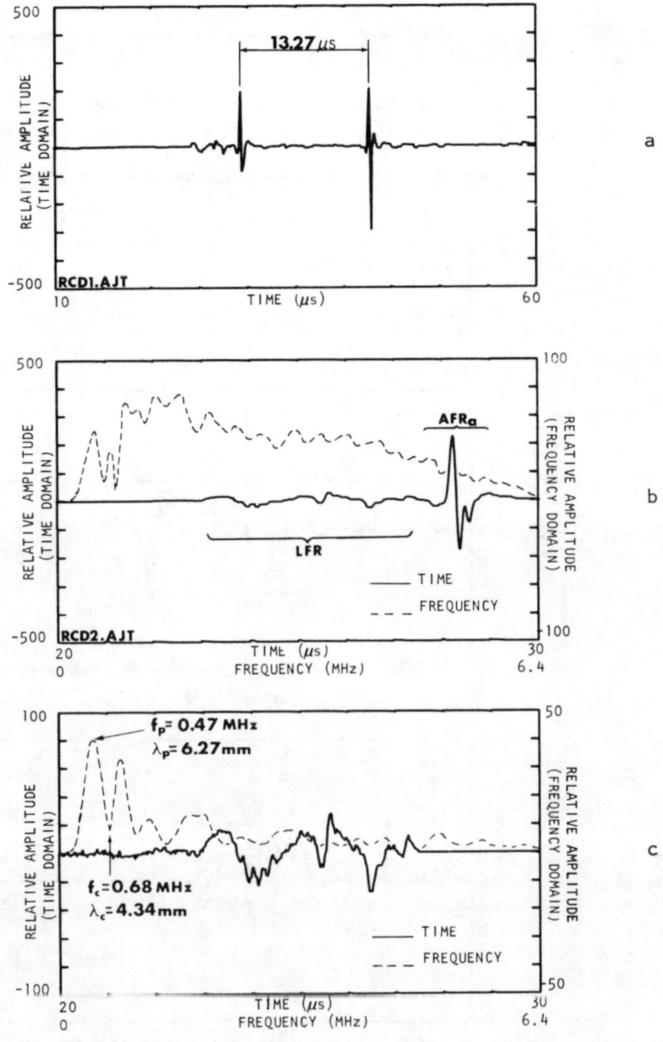

Figure 7. Sequence of samples of polluted fatigue crack under a 4000 lb flexural load
a. Full view of transmitted signals
b. Transmitted wave group and FFT
c. Leading wave group and FFT

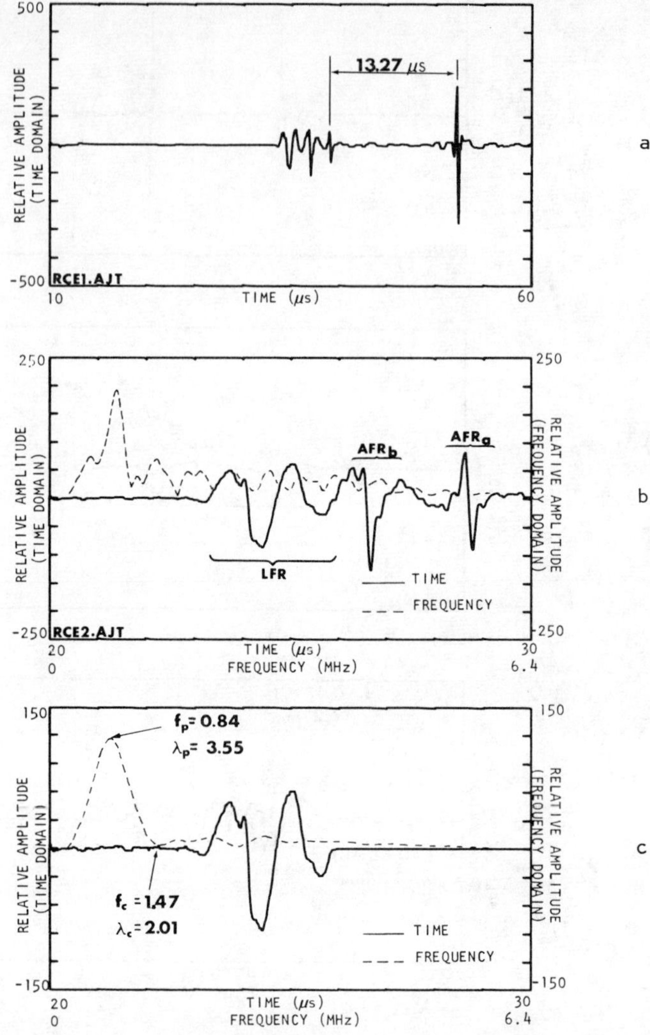

Figure 8. Sequence of samples of clean fatigue crack under no load
a. Full view of transmitted signals
b. Transmitted wave group and FFT
c. Leading wave group and FFT

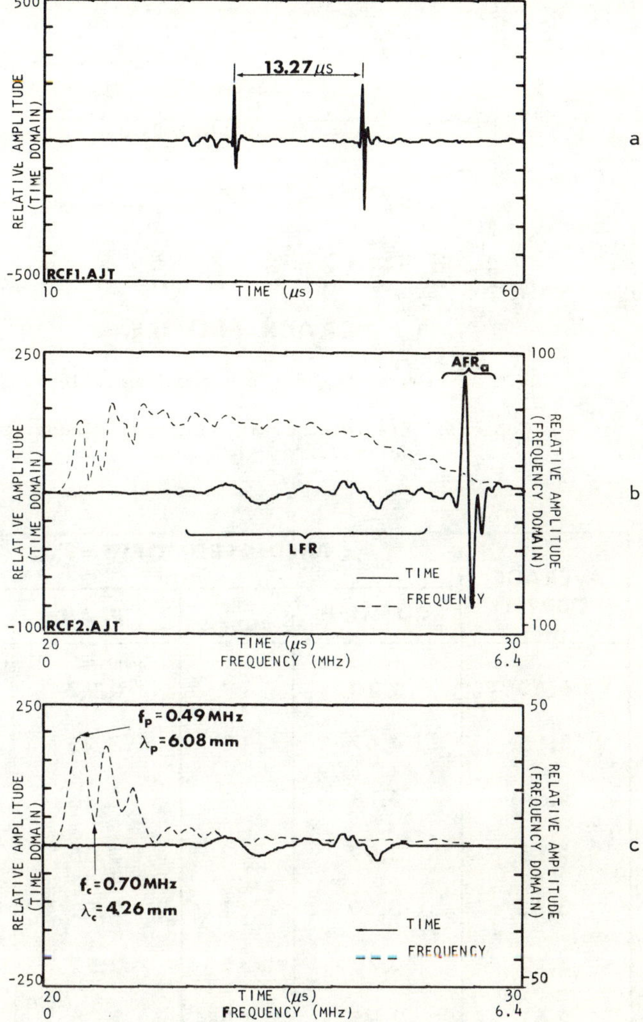

Figure 9. Sequence of samples of clean fatigue crack under an 8000 lb flexural load
a. Full view of transmitted signals
b. Transmitted wave group and FFT
c. Leading wave group and FFT

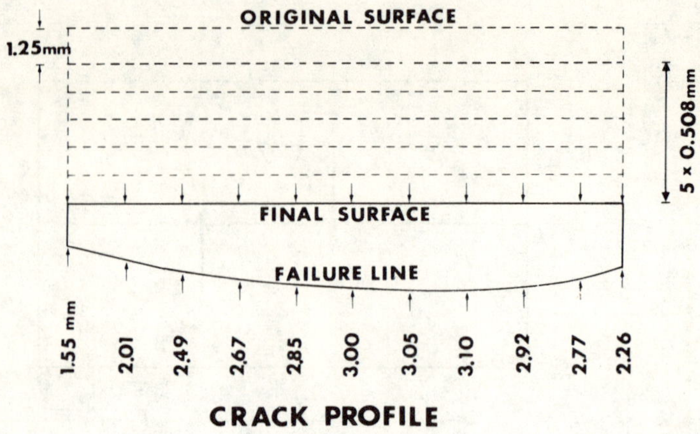

Figure 10. Fatigue crack profile

Table I. Predicted crack depths from equations derived from slot data

AVERAGE DEPTH mm	PREDICTED DEPTHS mm			
	CUT OFF	% ERROR	PEAK	% ERROR
6.40	5.20	+ 19	4.22	34
5.15	4.47	+ 13	3.25	37
4.64	4.18	+ 10	3.36	28
4.13	3.99	+ 3	3.05	26
3.63	3.60	- 1	2.52	31
3.12	3.28	- 5	2.52	19
2.61	2.95	- 13	2.25	14

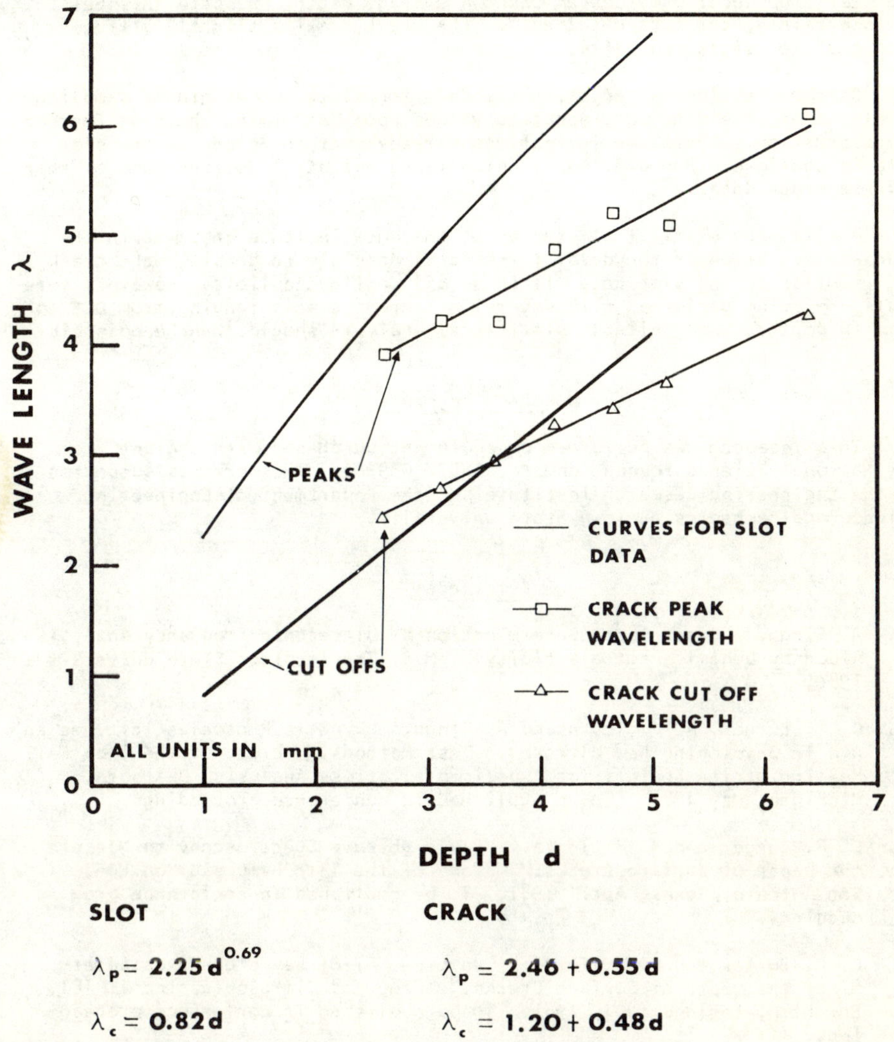

SLOT

$$\lambda_p = 2.25\,d^{0.69}$$

$$\lambda_c = 0.82\,d$$

CRACK

$$\lambda_p = 2.46 + 0.55\,d$$

$$\lambda_c = 1.20 + 0.48\,d$$

Figure 11. Slot and crack data results and equations

Conclusions

Actual cracks do not have flat profiles so that testing with a large diameter transducer provides an unknown average depth. Despite this built-in uncertainty, the data obtained for the crack remains slightly different from that for slots and slits.

Current testing is repeating the data for slots but at higher sampling levels, thus providing more accurate values upon performing the Fast Fourier Transformation. There are indications already that the slope of the slot data is shallower than previously established but still not the same as that for the crack data.

Preliminary statistical studies of the data indicate that within the standard deviation of the data, it is still possible to consider the crack data similar to the slot data within an 85% confidence limit. However, more tests are being performed with several different cracks ranging from 0.5 to 8 mm in depth. More reliable statistical analysis should then be possible.

Acknowledgments

This research was performed on equipment purchased with a grant from the National Science Foundation (No. ENG77-06970). The work was supported by the Engineering Research Institute and the Department of Engineering Science and Mechanics at Iowa State University.

References

1. A. Singh. "Crack Depth Determination by Ultrasonic Frequency Analysis Aided by Dynamic Photoelasticity." M.S. Thesis, Iowa State University, 1980.

2. C. P. Burger, A. J. Testa and A. Singh. "Dynamic Photoelasticity as an Aid in Developing New Ultrasonic Test Methods." Proc. of the 1981 Fall Meeting of the Society for Experimental Stress Analysis, Dearborn, Michigan, May 1981. To be published in conference proceedings.

3. C. P. Burger and A. J. Testa. "Rayleigh Wave Spectroscopy to Measure the Depth of Surface Cracks." Proc. of the 13th Symposium on NDE, San Antonio, Texas, April 1981. To be published in conference proceedings.

4. C. P. Burger and A. J. Testa. "On-Line FFT of Rayleigh Waves Determines the Depth of Surface Cracks." Proc. of Ultrasonics Int'l. '81, Brighton, England, July 1981. To be published in conference proceedings.

5. M. G. Silk. "The Determination of Crack Penetration Using Ultrasonic Surface Waves." _NDT Int'l._, 9(6) (December 1976), 290-297.

6. M. G. Silk. "Defect Sizing Using Ultrasonic Diffraction." _British J. of NDT_, 21(1) (January 1979), 12-15.

7. R. J. Hudgell, L. L. Morgan and R. F. Lumb. "Non-Destructive Measurement of the Depth of Surface Breaking Cracks Using Ultrasonic Rayleigh Waves." _British J. of NDT_, 16(15) (September 1974), 144-149.

8. M. V. Grigorev, A. K. Gurvich, V. V. Grebennikov and S. V. Semerkhanov. "Ultrasonic Technique for the Determination of Crack Dimensions." Soviet J. of NDT, 15(6) (June 1979), 495-500.

9. B. T. Khuri-Yakub and G. S. Kino. "Acoustic Surface Wave Measurements of Surface Cracks in Ceramics." J. Am. Ceram. Soc., 63(1-2) (January-February 1980), 65-71.

10. L. L. Morgan. "The Spectroscopic Determination of Surface Topography Using Acoustic Surface Waves." Acoustica, Vol. 30 (1974), 222-228.

11. K. G. Hall. "Crack Depth Measurements in Rail Steel by Rayleigh Waves Aided by Photoelastic Visualization." Non-Dest. Testing (London), 9(3) (June 1976), 121-126.

12. P. A. Doyle and C. M. Scab. "Crack Depth Measurement by Ultrasonics: A Review." Ultrasonics, 6(4) (July 1978), 164-170.

13. B. G. Yee, J. C. Couchman, J. W. Hagemayer and F. H. Chay. "Stress and the Ultrasonic Detection of Fatigue Cracks in Engineering Metals." Non-Dest. Testing, Vol. 7 (1974), 245.

14. D. M. Corbly, P. F. Packman and H. S. Pearson. "Accuracy Precision of Ultrasonic Shear Wave Flaw Measurements on a Function of Stress on the Flaw." Mat. Eval., Vol. 30 (1970), 103-110.

15. B. H. Lidington, D. H. Saunderson and M. G. Silk. "Interference Effects in the Reflection of Ultrasound from Shallow Slits." Non-Dest. Testing, Vol. 6 (1973), 185-190.

16. F. C. Cuozzo, E. L. Cambiaggio, J. P. Damiano and E. Rivier. "Influence of Elastic Properties on Rayleigh Wave Scattering by Normal Discontinuities." IEEE Trans. on Sonics and Ultrasonics, SV-24(4) (July 1977), 280-289.

17. S. Golan. "Optimization of the Crack Tip Ultrasonic Diffraction Technique for Sizing Cracks." Mat. Eval., Vol. 39 (February 1981), 166-169.

18. L. Kittineer. "Correction for Transducer Influence on Sound Velocity Measurements by the Pulse Echo Method." Ultrasonics, Vol. 15 (January 1977), 28-30.

19. A. Singh, C. P. Burger, L. W. Schmerr and L. W. Zachary. "Dynamic Photoelasticity as an Aid to Sizing Surface Cracks by Frequency Analysis." Mech. of Non-Dest. Testing, Ed. W. W. Stinchcomb, New York: Plenum Press (1980), 277-292.

20. L. W. Zachary, C. P. Burger, L. W. Schmerr and A. Singh. "Surface-Crack Characterization by Use of Dynamic Photoelasticity Spectroscopy." Paper presented at the SESA Spring Meeting, San Francisco, California, May 20-25, 1979.

21. A. Singh and C. P. Burger. "A New Technique for Determining the Depth of Surface Cracks by Rayleigh Waves." ASNT, Fall Conference, Houston, Texas, October 1980.

22. C. P. Burger and A. Singh. "An Ultrasonic Technique for Sizing Surface Cracks." Proc. DARPA/AF, Review of Progress in Quantitative NDE, San Diego, California, July 1980.

23. I. A. Viktorov. "Rayleigh and Lamb Waves." First ed. New York: Plenum Press, 1967.

24. J. D. Achenbach. "Wave Propagation in Elastic Solids." Vol. 16, London: North-Holland Publishing Co., 1973.

25. H. Kolsky. "Stress Waves in Solids." First ed. London: Claredon Press, 1953.

26. A. J. Testa. "Non-Destructive Measurement of Surface Cracks Using Ultrasonic Rayleigh Waves." Ph.D. Dissertation, Iowa State University, 1982.

THE DETECTION OF STRESS CORROSION CRACKING BY PRECISION, CONTINUOUS MODULUS MEASUREMENTS

Steve H. Carpenter

Department of Physics
University of Denver
Denver, Colorado 80208
U.S.A.

The use of precision, continuous modulus measurements to detect and monitor the nucleation and growth of cracks during stress corrosion will be reported on. The nucleation of cracks within a sample should cause a decrease in the elastic modulus of the material. Accurate continuous measurements of the Young's modulus have been made using a modified composite oscillator. Using a specially designed test sample, modulus measurements were carried out on 304 stainless steel while subjected to different environmental solutions while under stress. Results indicate a good correlation between stress corrosion cracking and a measured decrease in the modulus. The technique shows promise of providing a rapid and reliable method of determining the susceptibility of materials to stress corrosion cracking. In addition, data on modulus changes in pure iron as a result of cathodic charging with hydrogen will be presented.

Introduction

The basic premise of the investigation reported on in this paper is that a sample of a given material containing cracks should have a lower elastic modulus than an identical sample of the same material containing no cracks. Bristow (1) has examined this question in detail and has shown for a sample of volume V containing N randomly oriented penny shaped cracks of radius a, the change in the Young's modulus is given by

$$Yeff = Yo \left[1 - \frac{16(1-\nu^2)(10-3\nu)}{45(2-\nu)} \frac{N}{V} a^3\right] \tag{1}$$

where Yo is the Young's modulus of the material with no cracks and ν is Poisson's ratio. With a Poisson's ratio of 0.3, Equation (1) reduces to

$$Yeff = Yo \left[1-1.73 \frac{N}{V} a^3\right] \quad . \tag{2}$$

Clearly a decrease in the Young's should be observed as cracks nucleate and grow in the test material.

The modulus of samples of 304 stainless steel was monitored continuously after the samples were strained and maintained under load at 7% and exposed to different environmental solutions. Modulus measurements were also carried out on both 304 stainless steel and pure iron samples that were cathodically charged under zero stress (2). In both cases the samples were examined metallographically for evidence of cracking. Results show good correlation between a decrease in the modulus and the occurrence of cracking.

Experimental

The Young's modulus of the test specimens was continuously monitored by accurately measuring the resonant frequency of the specimen while driving it in a standing longitudinal wave configuration. A modified Marx composite piezoelectric oscillator (3) was used to drive the sample at a resonant frequency of approximately 50 kilohertz. The resonant frequency was measured to within one Hertz. A closed-loop crystal driver continuously maintained the sample at its resonant frequency at a predetermined (but adjustable) vibratory strain amplitude. A microprocessor, interfaced with the closed-loop crystal driver, gave continuous digital and analog printout of the resonant frequency. The resonant frequency of the total wave train is the measured quantity from which the resonant frequency of the sample is calculated. Once the resonant frequency of the sample is known the Young's modulus can easily be calculated.

For the cathodic charging experiments the samples were rods whose length was carefully cut to be exactly one-half wavelength (approximately 5.1×10^{-2} m), the center of the sample being a displacement node. Hence, the necessary electrical connection needed for cathodic charging could be made at the center of the sample without interfering with the standing wave. A schematic of the apparatus is shown in Fig. 1. For more details see reference 2. For the stress corrosion experiments, a specially designed sample, as shown in Fig. 2, was used. This type of sample allows for continuous modulus measurements while simultaneously applying a load and exposing the sample to a corrosive test solution. The sample, as shown, is three half wave lengths long; the gripping nodes and the solution cup being silver soldered on at displacement nodes do not interfere with the standing waves. The complete experimental apparatus is shown schematically in Fig. 3.

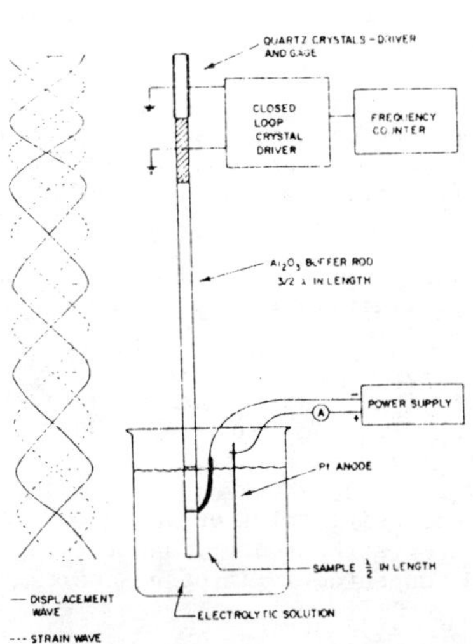

Figure 1. Schematic of experimental apparatus used to measure the modulus while cathodically charging the test sample.

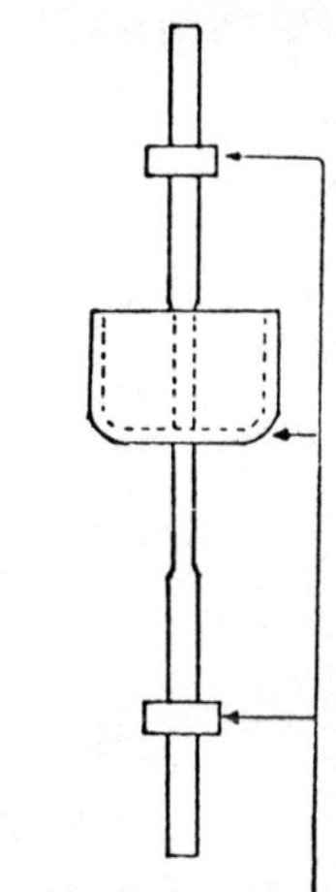

Figure 2. Schematic of special sample used to measure modulus while stress corrosion cracking is occurring.

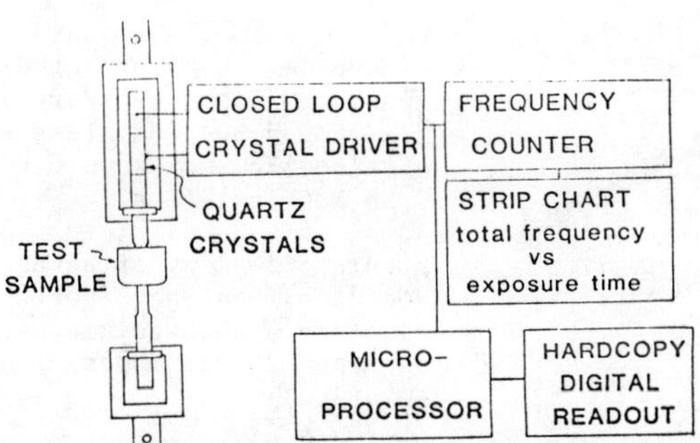

Figure 3. Schematic of experimental apparatus used to measure the modulus while stress corrosion is cracking the sample.

Results and Discussion

Stress Corrosion Studies

Experiments to investigate the use of continuous modulus measurements to probe stress corrosion cracking have been carried out in the following manner:

1. Specially designed samples (Fig. 2) were deformed in tension 7% total strain. The samples were designed and fabricated such that the 7% strain produced a length giving a resonant frequency which matched that of the driver and gauge quartz crystals.

2. After deforming to 7% strain the samples were allowed to stress relax until there was no measurable change in applied load during a time segment of approximately one hour.

3. Upon complete stress relaxation a corrosive solution was added to the solution cup on the sample and the resonant frequency was continuously monitored for an extended time period, usually around 10^5 seconds.

4. Following the long-term exposure the corrosive solution was removed and the samples were examined for evidence of stress corrosion cracking. Some samples were unloaded, sectioned and inspected while others were pulled to failure and then examined.

Figures 4a, 4b, and 4c show actual data, i.e., resonant frequency vs. exposure time, from strip chart traces for samples exposed to different corrosive solutions. The data shown were recorded at room temperature well into the test. Figures 4b and 4c show little or no consistent change in the measured resonant frequency for samples exposed to either Na_2SO_4 or Na_2SO_4/H_2SO_4. However, as shown in Fig. 4a there is a significant and continuous linear decrease of the resonant frequency of samples exposed to H_2SO_4/ NaCℓ.

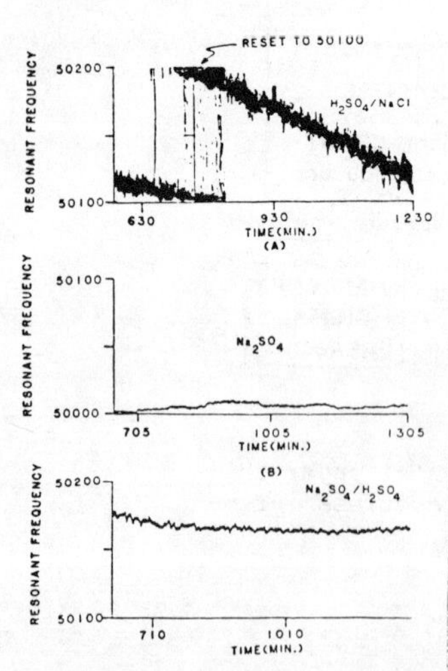

Figure 4a. Copy of actual data trace of resonant frequency (Hertz) vs. exposure time for a 304 stainless steel specimen, strained 7% and exposed to a 5N H_2SO_4/ 0.5N NaCℓ solution.

Figure 4b. Copy of actual data trace of resonant frequency (Hertz) vs. exposure time for a 304 stainless steel specimen, strained 7% and exposed to a 5N Na_2SO_4 solution.

Figure 4c. Copy of actual data trace of resonant frequency (Hertz) vs. exposure time for a 304 stainless steel specimen, strained 7% and exposed to a 5N Na_2SO_4/ 5N H_2SO_4 solution.

After exposure of approximately 10^5 seconds the corrosive solution was removed and the samples were either unloaded and examined or were pulled to failure and examined for evidence of stress corrosion cracking. As expected, samples exposed to Na_2SO_4 and Na_2SO_4/H_2SO_4 showed no evidence of cracking and exhibited the standard ductile behavior common to 304 stainless steel. However, samples exposed to $H_2SO_4/NaC\ell$ showed substantial cracking. Figure 5a shows the fracture surface of one of these samples. The surface cracking is clearly evident. Figure 5b is a photomicrograph of a cross-section of the sample showing typical branching type transgranular chloride stress corrosion cracking. The $H_2SO_4/NaC\ell$ solution is so aggressive to 304 stainless steel that even at room temperature many of the cracks have been overcome by environmental dissolution.

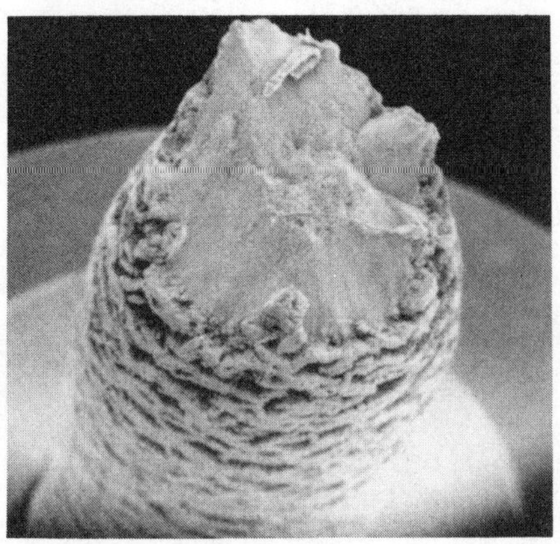

Figure 5a. 304 stainless steel specimen showing necked down area and environmental cracking and final mechanically induced fracture in center. SEM 30X

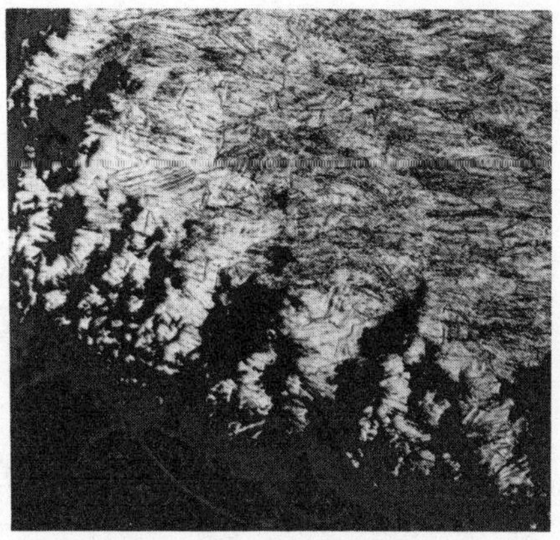

Figure 5b. Photomicrograph of cross section of coupon showing typical branching type transgranular chloride stress corrosion cracking. $H_2SO_4/NaC\ell$ is so aggressive to 304 s.s. that even at room temperature many of the SCC's have been overcome by environmental dissolution. Mag. 100X Etch Oxalic electrolytic.

As with any technique it is necessary to determine what effect the method of measurement has on the observed data. In this case it is necessary to establish the effect the continuous vibratory strain imposed by the quartz crystals, i.e., is the cracking due to or accelerated by the imposed vibrations necessary for measurement of the resonant frequency.

To answer these questions the following experiment was carried out. Instead of driving the sample continuously the sample was driven only for the time required to make a measurement, approximately two minutes for each data point. Results of this experiment are plotted in Fig. 6 along with data from a continuous experiment. A comparison of the data indicates that the continuous application of the vibratory strain does accelerate the process but only by a slight amount. The same general behavior is observed but with a small time shift.

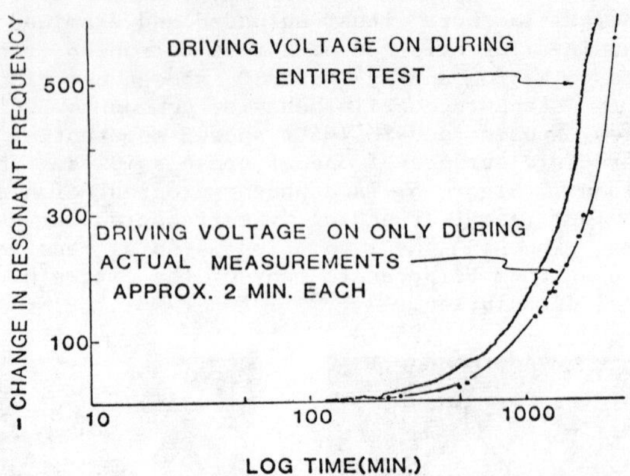

Figure 6. Negative change in resonant frequency (Hertz) vs. log exposure time for a 304 stainless steel specimen, strained 7% and exposed to a 5N H_2SO_4/0.5N NaCℓ solution. First curve for continuous application of driving voltage. Second curve driving voltage applied only during actual measurement. Approximately 2 minutes were required for each data point.

Preliminary attempts have been carried out to determine the sensitivity of the modulus measurement technique. A series of tests were carried out using corrosive solutions of 5 normal H_2SO_4 mixed with an NaCℓ solution of various normalities. Figure 7 shows a portion of the data generated during this series of tests. The data are given as the negative change in resonant frequency as a function of exposure time. The data for the 0.5 normal NaCℓ solution and the 0.3 normal NaCℓ solution are essentially identical. The data can be described as having an incubation time of approximately 10^4 seconds after which the resonant frequency decreases in a linear function of exposure time. A reduction of the Cℓ ion concentration using a 0.1 normal NaCℓ solution causes a noticeable change in the data. With the 0.1 normal NaCℓ solution there is a considerably longer incubation time before a decrease in the resonant frequency is observed. However, once the resonant frequency begins to decrease, the rate of decrease is approximately linear and at the same rate is observed for the other solutions. The fluctuations and deviations in the linear portions of the data in Fig. 7 are believed to be due to temperature variations during the long time duration of the tests. A further reduction of the Cℓ ion concentration using a 0.05 normal solution gives a corrosive solution which causes no measurable change in resonant frequency during long time exposure.

Figure 8 shows micrographs at three magnifications from the electron scanning microscope for a sample exposed for 1.6×10^5 seconds in a solution of 5 normal H_2SO_4 plus 0.1 normal NaCℓ. The sample was simply unloaded after the corrosive solution was removed. The pitting of the surface is obvious in the lower magnifications while cracking is clear in the highest magnification. By comparison Fig. 9 shows micrographs of a sample tested in the 0.05 normal NaCℓ solution where no change in resonant frequency was observed. The differences in the two sets of micrographs are clear.

As stated in the introduction, one purpose of this investigation was to determine if the technique of modulus measurements could be used to rapidly and reliably determine if test materials were susceptible to stress corrosion cracking in specific environments. Although more tests are required, the data to date indicate that the technique has great promise. The impressive feature is the sensitivity of the technique to the cracking process. In all of the samples exposed to the solution, indications of cracking were evident after 4-5 hours of testing. This is a significantly faster technique than any available today.

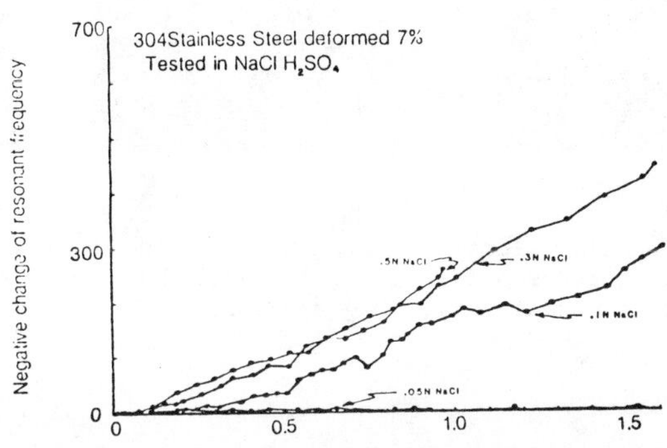

Figure 7. Negative change in resonant frequency vs. linear exposure time for 304 stainless steel specimens strained 7% and exposed to different H_2SO_4/NaCℓ solutions.

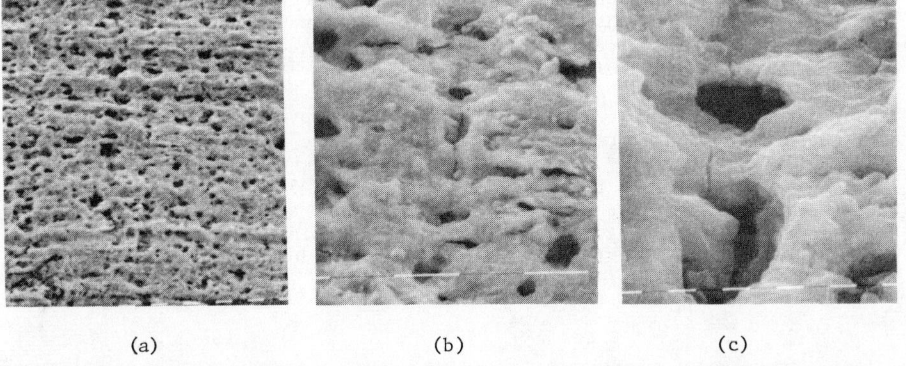

(a) (b) (c)

Figure 8. Scanning electron micrographs of the surface of a sample of 304 stainless steel exposed for 1.6×10^5 seconds in a solution of 5 normal H_2SO_4 plus 0.1 normal NaCℓ. Mag a) 320X, b) 1250X, c) 5000X

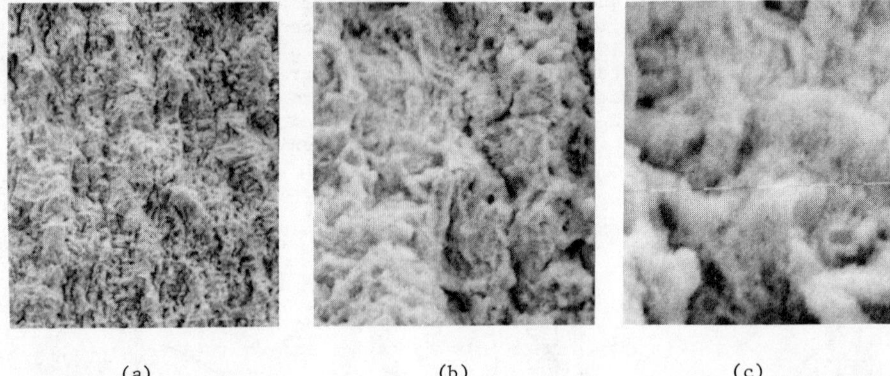

(a) (b) (c)

Figure 9. Scanning electron micrographs of the surface of a sample of 304 stainless steel exposed for 1.6×10^5 seconds in a solution of 5 normal H_2SO_4 plus 0.05 normal NaCℓ. Mag a) 320X, b) 1250X, c) 5000X

Cathodic Charging

Experiments were also conducted using the apparatus shown in Fig. 1 to determine if modulus measurements were sensitive to cracking induced by cathodic charging. Figure 10 shows actual data, i.e. the change of the resonant frequency of the sample as a function of charging time. The main coordinates in Fig. 4 show data as a function of time on a log scale while the insert shows data as a function of time on a linear scale. The data were obtained at a constant temperature of $0^\circ C \pm 0.5^\circ C$. The pure iron samples (Grade 1) were annealed at $700^\circ C$ for 4 hours in a vacuum before charging. An examination of the data in Fig. 4 shows that for pure iron there was very little change in the resonant frequency until after 10^2 seconds of charging, after which the resonant frequency, and hence the modulus, decreased steadily with further charging. (Note that the negative frequency change is the parameter plotted.) Data shown in the insert of Fig. 4 indicate that the resonant frequency decreased rapidly and then leveled off to some plateau value. Metallographic examination indeed confirmed cracks, primarily along grain boundaries, which developed during charging. No change in the resonant frequency of 304 stainless steel was observed on charging for up to 10^5 seconds. No cracks could be found by metallographic examination of the charged 304 stainless steel samples. For more details concerning modulus measurements during cathodic charging the reader is referred to reference 2.

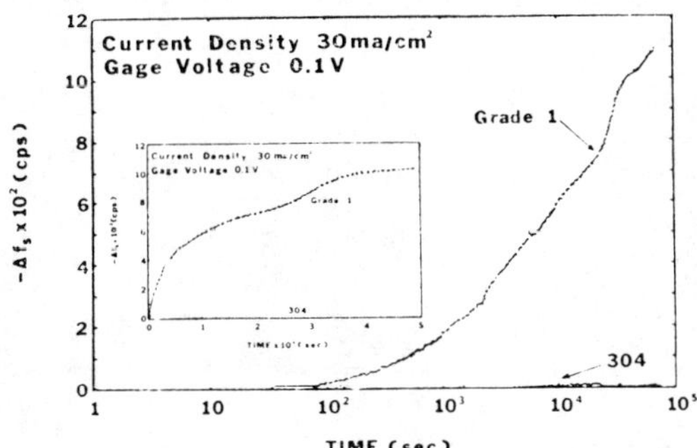

Figure 10. Negative change of the resonant frequency of test samples as a function of charging time. Major coordinates show charging time on a log scale. Insert shows charging time on a linear scale.

Conclusions

The preliminary results reported in this paper indicate that continuous precision modulus measurements is a promising technique to detect and study stress corrosion cracking. While no definitive relationship between the amount of cracking and the measured change in modulus was obtained, good correlation between the occurrence of cracking and a negative change in resonant frequency was observed. The imposition of the continuous vibratory strain necessary to measure the resonant frequency was shown not to influence the data seriously. The technique, at least for the materials and solutions tested, gave indications that cracking was occurring significantly faster than any technique available today. While more work is necessary, the promise of the technique is clearly evident.

Acknowledgments

Acknowledgment and thanks are extended to Mr. O. Siebert of Monsanto Company, St. Louis, MO for helpful discussions and partial financial support of this work. Financial assistance was also received from the National Science Foundation, DMR Grant 77-23782 and from a National Association of Corrosion Engineers Seed Grant.

References

1. J.R. Bristow, British J. of Appl. Phy. 11, p. 81 (1960).

2. S.H. Carpenter and J.E. Fawks, Scripta Metall. 15, p.699 (1981).

3. J. Marx, Rev. Sci. Instrum, 22, p. 503 (1951).

ACOUSTIC HARMONIC GENERATION FOR REMAINING FATIGUE LIFETIME ESTIMATION IN Ti 6Al-4V ALLOY

W.L. Morris, R.V. Inman, and W.J. Pardee

Rockwell International Science Center
Thousand Oaks, CA 91360
USA

Second harmonic acoustic signals produced when fatigue specimens of Ti 6Al-4V are exposed to 5 MHz fundmental surface acoustic waves are measured at intervals during fatigue. The enhanced harmonic signal that develop in the fatigued material is attributed to the anelastic opening and closing of small surface microcracks in the fundamental acoustic wave. A model of the generation process is proposed which hypothesizes for titanium that the preponderance of the harmonic signals are associated with grain sized crystallographic cracks. The model is evaluated by comparing predicted and measured harmonic amplitudes developed by fatigue for the titanium alloy in two grain sizes. Concurrence between theory and experiment is achieved if it is assumed that the individual contributions of each crack sum coherently to produce the total received harmonic signal. The potential application of the harmonic technique is to estimate the average remaining lifetime of titanium components subjected to low cycle fatigue.

Introduction

Acoustic waves at harmonic frequencies are generated as fundamental acoustic waves traverse a solid. Lattice anharmonicity was the first source of harmonic generation identified (1-3). Thereafter, Hikata et al. (4) demonstrated that dislocation motion in an acoustic wave can also contribute to harmonic generation. Subsequently, many investigators (5-11) have studied the complex interaction between fatigue and deformation, applied stress and alloy heat treatment and the second-harmonic signals developed in steel, titanium, and aluminum. Many of these results could be interpreted by modeling the effect of these variables on dislocation mobility and loop lengths. More recent studies have focused on the role of fatigue cracks in the production of harmonic waves (12-15). The anharmonicity apparently arises as a crack opens and closes along a portion of the crack front as an acoustic wave passes normal to fracture plane. Theoretical (12) considerations and experimental studies (13) demonstrate that a small contact stress normal to the crack face, in the range of 0.05-0.5 MPa, produces the maximum harmonic generation efficiency. This is a contact stress comparable with the stress amplitude developed in a solid by typical acoustic transducers.

An important potential application of acoustic harmonics is to estimate the average remaining fatigue lifetime of a structure by characterization of the fatigue damage it has sustained. To be practical, a substantial harmonic signal must be generated from the numerous small microcracks in the surface, that are the common precursors to low cycle fatigue failure. Buck et al. (15) have demonstrated for Al 7075-T6 and Morris et al. (14) for Al 2219-T851 that remaining fatigue lifetime can be estimated by using surface waves to detect harmonic signals produced by small surface cracks. For the same microcracking density on the surface, the harmonic amplitude is larger in 7075. Buck et al. (15) have hypothesized that this is a consequence of a larger ratio of crystallographic to non-crystallographic cracks in the 7075 alloy. Crystallographic cracks close more uniformly than their non-crystallographic counterparts and should allow small contact stresses to be developed over a larger area within each microcrack.

Recent studies (16) show that there is a considerable variation in the character of closure of small surface cracks with alloy grain size and yield strength. Unlike aluminum, short cracks in high strength alloys tend to be held open more by asperities. Also, the residual openings increase with grain size. In this paper, the effect of grain size in a high strength Ti 6Al-4V alloy on harmonic amplitude is examined and compared to reported results for aluminum. The hypothesis that the harmonics are preferentially generated by crystallographic cracks is evaluated by comparing measured amplitudes to those predicted by a model of the harmonic generation process derived from that assumption.

Experimental Procedures

Ti 6Al-4V plate of 0.18 cm thickness was obtained in the mill annealed condition, and had a yield strength of 1000 MPa. Its microstructure consisted of 4 μm diameter primary α grains separated by a fine distribution of transformed β phase. A portion of this plate was heat treated for 8 hours at 300°C and oil quenched to produce α grains 8 μm in diameter. Tapered cantilever beam specimens, such as illustrated in Fig. 1, were machined from these materials. The surface was prepared to minimize residual surface stresses by polishing with alumina powder to remove 0.007 cm of material. Success of this procedure was confirmed by x-ray diffraction analysis. Specimens were fatigued in bending in laboratory air and at intervals the loading was interrupted and the second harmonic amplitude was measured as a function of applied surface stress, and of the amplitude of the fundamental wave.

The acoustic transducers and signal processing used for this study have been described elsewhere (14). Briefly a fundamental acoustic frequency of 5 MHz was generated using a slab of PZT mounted on a fused quartz wedge. A water wedge, illustrated in Fig. 1, was used to couple the acoustic pulse to a Rayleigh wave in the top surface of the specimen. A similar transducer comprised of 10 MHz PZT and mounted at the opposite end of the specimen received both the harmonic and fundamental signals. The transmitting and receiving transducer elements were separated by approximately 7 cm. Cracks were initiated within a strip of material approximately 3 cm long located midway between the transducers. Filters placed in the output of the transducer were tuned to either 5 or 10 MHz depending upon which signal was to be measured. Received signals were further processed using a heterodyne receiver and displayed on an oscilloscope for measurement of the amplitude. The system was operated in the pulse mode by exciting the 5 MHz PZT with 200 volt peak-to-peak 5 MHz radio frequency signals of approximately 20 cycles duration. Care was taken in the electronic design and qualification to

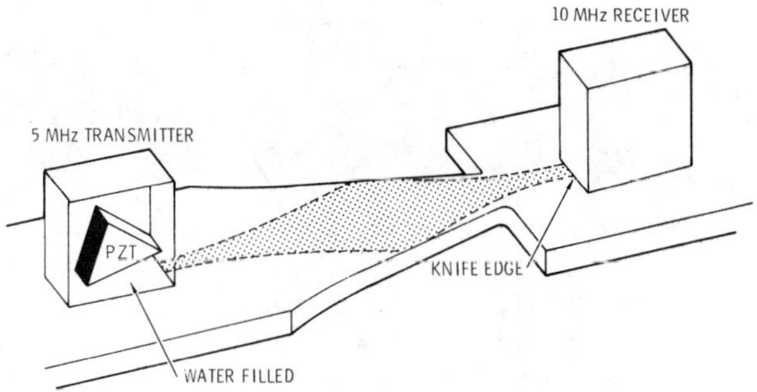

Figure 1–Tapered cantilever beam fatigue specimen used for harmonic generation study. Acoustic transducers consisting of 5 or 10 MHz PZT mounted on quartz wedges are used to launch and receive Raleigh waves on the specimen surface through water wedges mounted on either end of the specimen.

ensure that the background harmonic signals received from unfatigued specimens were actually produced in the titanium. Spurious signals must always be of concern in harmonic generation experiments as they can arise from unexpected sources. For instance, it was discovered that conventional lucite wedge transducers commonly used to couple Raleigh waves to titanium could not be used because more harmonics were developed in the wedges than in the titanium specimen itself.

For selected specimens the numbers and lengths of cracks developed in the gauge section were determined at intervals during fatigue. These measurements were done using an optical microscope which utilized a loading jig to place the surface of the specimens in tension. This opened the small surface cracks for greater visability.

Results

The titanium specimens were fatigued at a peak cyclic stress amplitude of approximately 1000 MPa and at a minimum/maximum load ratio of -0.5. This resulted in specimen failure in 1500 to 4000 cycles depending upon the material and precise cyclic stress amplitude used. A quadratic relationship between the second harmonic amplitude (A_2) and the fundamental amplitude (A_1) was found prior to fatigue of specimens of both heat treatments (i.e., $A_2 \propto A_1^2$, Fig. 2). This relationship is expected if the principal source of harmonics is from dislocation motion in the acoustic wave (15). To partially compensate for the variation in second harmonic amplitude with A_1, data presented below are normalized by dividing the received harmonic amplitude by the square of the received fundamental amplitude.

4 μm Diameter α Phase Material

Cracks in the 4 μm diameter α phase material initiated in fatigue bands within the α phase. The harmonic amplitude rose abruptly during the first few hundred cycles and then decreased slowly with fatigue until it returned to the unfatigued background level at 50% of the average expended fatigue lifetime (Fig. 3). As has been reported previously for aluminum alloys, elevated levels of harmonic generation above the prefatigue background were associated with a departure from the quadratic relationship

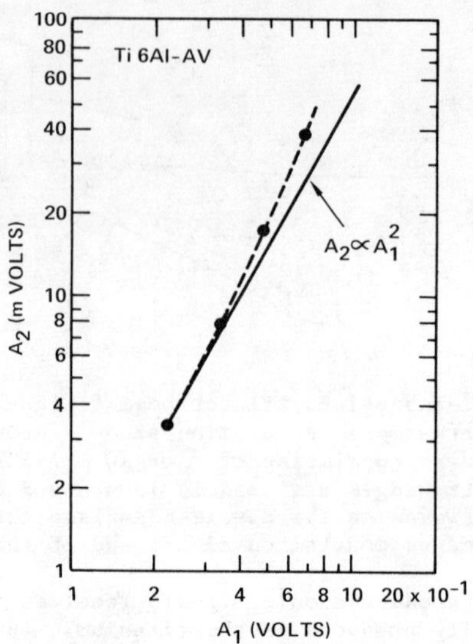

Figure 2 - Received second harmonic amplitude A_2 versus received fundamental amplitude A_1. Solid line dependence taken before fatigue is apparently due to dislocation sources. Dashed line taken after fatigue is typical of the change due to surface microcracking.

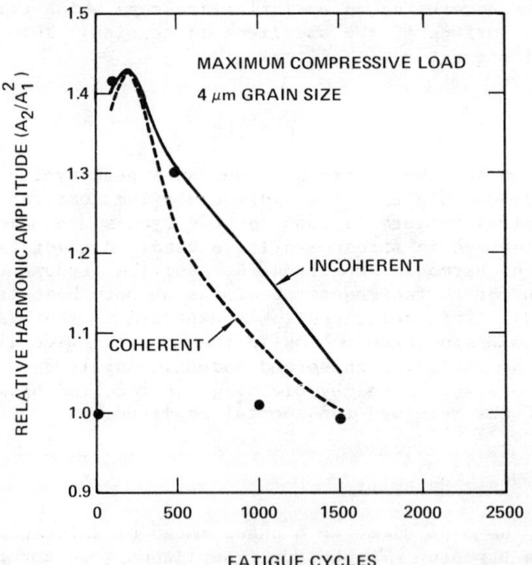

Figure 3 - Progressive change in normalized harmonic amplitude with fatigue for the 4 μm a phase alloy. Dashed line from incoherent model. Solid line from coherent generation model.

between A_2 and A_1. An example is shown in Fig. 2 and is consistent with the production of harmonic energy by the opening of microcracks in the acoustic wave. The maximum fatigue induced harmonic amplitudes were detected when the applied stress was in the range of -500 to 0 MPa. Later during fatigue, as longer cracks began to develop in the gauge section, both the received fundamental and harmonic amplitudes were attenuated if the applied stress was tensile, due to opening of the surface microcracks. The curves in Fig. 3 are derived from a model discussed later.

8 μm Diameter α Phase Material

Cracks in the 8 μm diameter material initiated at interfaces between the α and β phases. Harmonic generation increased with progressive fatigue (Fig. 4) and the maximum harmonic amplitude was developed with the specimens under tensile load (Fig. 5). Predicted values given in Fig. 5 were obtained using the model discussed below.

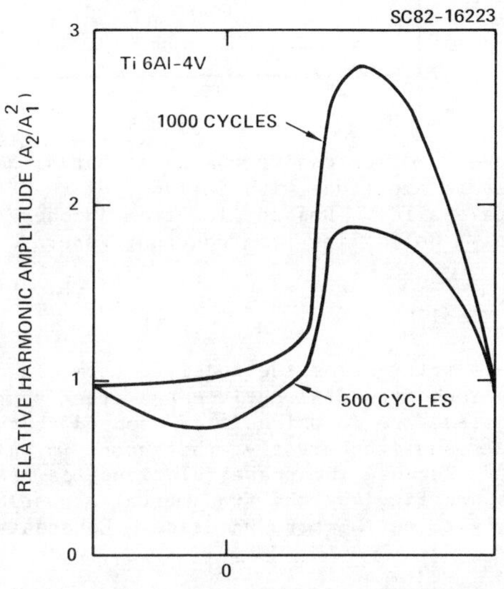

Figure 4 - Variation in received normalized harmonic amplitude as a function of applied surface stress for two intervals in fatigue. Data for the 8 μm α phase alloy.

Discussion

A simple model is proposed to relate the received harmonic amplitude to the number and characteristics of cracks in the portion of the specimen exposed to 5 MHz acoustic waves. It is assumed that crystallographic cracks produce the preponderance of the harmonic signals. Satisfactory agreement is achieved with measured harmonic amplitudes during fatigue for both alloy grain sizes.

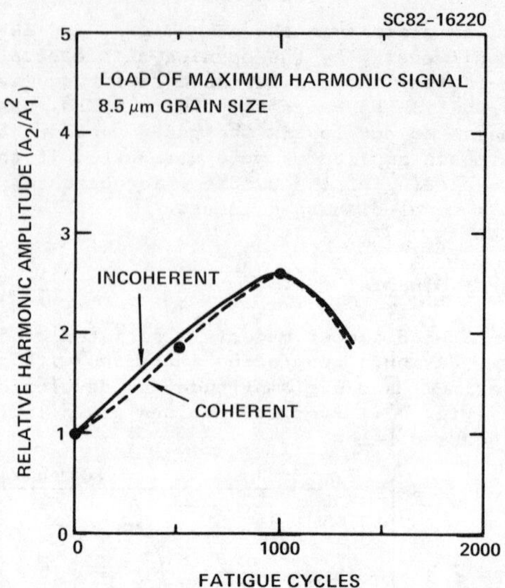

Figure 5 - Progressive change in normalized harmonic amplitude with fatigue for the 8 μm α phase alloy. Dashed line from incoherent model. Solid line from coherent generation model.

Harmonic Generation Model

An expression is written for the received harmonic amplitude (A_2) by summing the contributions from dislocations and crack sources. Two forms of the model are considered, one in which it is assumed that the contributions from individual cracks sum incoherently, the second in which the contributions add coherently. Because the phase relationships between the individual harmonic signals are fixed by the fundamental acoustic wave, the later presentation is likely to be the more accurate. Coherency is considered in more detail later.

For the incoherent case we write

$$A_2 = \sqrt{a_d^2 + a_c^2 \cdot \chi(N)} \quad . \tag{1}$$

a_d is the integrated harmonic amplitude from all dislocation surfaces. a_c is the harmonic amplitude produced by a single crystallographic crack. a_c is taken to be a constant, as the approximation is made that all crystallographic cracks have the same length equal to the mean grain size. The rationale for this model is that early crack growth to grain size in both alloys is rapid and likely to be crystallographic or interfacial. Crack growth into the second grain is frequently accompanied by fracture surface roughness which tends to hold the crack mouth open. $\chi(N)$ is the number of grain sized crystallographic cracks in the gauge section after N fatigue cycles. A normalized fundamental amplitude $A_1 = 1$, is assumed. Before fatigue $\chi(0) = 0$, and substitution of these conditions into Eq. (1) gives

$$A_2/A_1^2 = \sqrt{1 + a_c^2 \cdot \chi(N)} \quad . \qquad (2)$$

Similarly the coherence assumption gives

$$A_2/A_1^2 = \sqrt{1 + (a_c \cdot \chi(N))^2} \quad . \qquad (3)$$

Both A_1 and A_2 change progressively as the acoustic wave traverses the specimen gauge section. Morris et al. (14) have discussed elsewhere the analysis of harmonic data to extract an accurate A_1 to A_2 coupling coefficient in such a circumstance. In prior studies using aluminum alloys, fatigue induced changes in attenuation of both A1 and A2 were progressive and large making such calculations mandatory for complete understanding of the harmonic phenomenon. In this study the cracking density is low, and the attenuation of A_1 is nearly constant during fatigue, except near specimen failure. Fundamental insight into the nature of harmonic generation source can, therefore, be obtained by considering only the received amplitudes of A_1 and A_2. This is the approach taken in this paper.

Model Implementation

Both models were evaluated by determining $\chi(N)$ experimentally. A single value of a_c was then chosen for each material to give the theoretical curves in Figs. 3 and 5. To obtain $\chi(N)$, first a histogram $\Sigma(N_o,2c)$ (Fig. 6) giving the number of cracks/cm^2 per unit crack length was determined after N_o fatigue cycles on specimens approaching failure. Variable 2c is the crack length. $\chi(N)$ was then estimated for $N < N_o$ by calculating, for selected values of N, the numbers of grain sized cracks present at N cycles required to produce the distribution $\Sigma(N_o,2c)$ after (N_o-N) fatigue cycles. To make this calculation a growth rate (dc/dN) law of the form

$$\frac{dc}{dN} = Ac \quad , \qquad (4)$$

was assumed. This expression has been rationalized from short crack growth rate data in an aluminum alloy (17). Variable c is the half crack length at the surface. Coefficient A was determined from average growth rates measured for the two titanium alloys. Furthermore, it was assumed that the initial size of all the cracks corresponded to the mean grain size, and that each crack took an average of 200 loading cycles to begin to propagate into the next grain beyond that of initiation.

Further Assessment of Coherency

If the microcracks were all located on a line connecting the transmitting and receiving transducers, the harmonic amplitudes developed by each crack would (neglecting attenuation) add coherently to produce the received amplitude. The cracks are actually distributed across a cylindrically expanding fundamental incident wave front, and the received harmonic surface wave is, itself, a cylindrical wave. The geometric effects produce some loss of coherence, by an amount that can be estimated by an extension of Richardson's model (12). The harmonics are generated when opposing fracture surfaces touch, with one surface acting as a driving stress on the

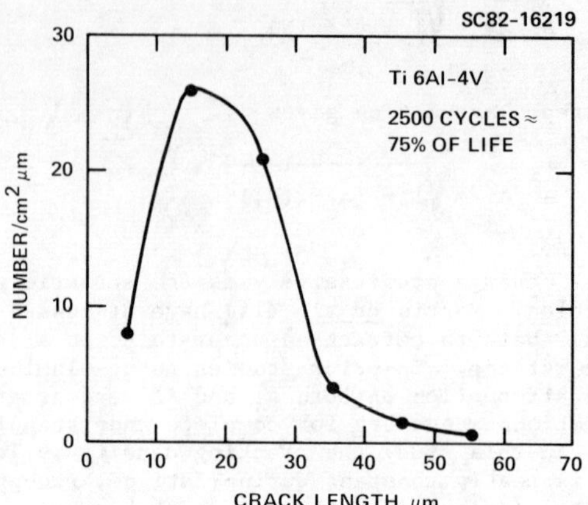

Figure 6 - Smoothed histogram of the numbers of cracks/cm^2 versus crack length in the gauge section of the 8 μm grain size material after 2000 fatigue cycles.

other. Let the nth crack be located at $\underset{\sim}{r}_n = (x_n, y_n)$, and the incident surface wave be presented (18) by the displacement function $u_i(t, \underset{\sim}{x})$ with

$$u_r(t,\underset{\sim}{r}) = -2A_0 H_1^{(1)}(kr)\, e^{-i\omega t}\, [(\beta^2 + k^2)e^{\alpha z} - 2\alpha\beta\, e^{\beta z}] \quad , \tag{5}$$

$$u_\theta = 0 \quad , \tag{6}$$

and

$$u_z = 2\alpha A_0 H_0^{(1)}(kr)\, e^{-i\omega t}\, [(\beta^2 + k^2)e^{\alpha z} - 2k^2 e^{\beta z}] \quad , \tag{7}$$

where

$$\alpha = (k^2 - \omega^2/c_L{}^2)^{1/2} \tag{8}$$

$$\beta = (k^2 - \omega^7/c_T{}^2)^{1/2} \tag{9}$$

$$k = \omega/c_R, \text{ and} \tag{10}$$

c_L, c_T and c_R are, respectively, the longitudinal, transverse, and Rayleigh velocities, and H_n^1 is the Hankel function of order n. This is the simplest possible angular form of a cylindrically expanding wave. More general angular dependence is described elsewhere (18).

The driving force for harmonic generation is the stress t on the receiving side of the microcrack,

$$t = C\,\varepsilon_{xx}\,\theta\,(C\,\varepsilon_{xx} - \sigma_{cc}) \tag{11}$$

where C is an appropriate elastic modulus and σ_{cc} is the stress required to close the crack. ϕ is a unit step function and is zero if its argument is $\leqslant 0$, and is otherwise unity. ε_{xx} is a displacement parallel to the line between the transmitting and receiving transducers (e.g., approximately perpendicular to the crack face). The component t_2 of t with time dependence $\exp(-2i\omega t)$ can be easily shown to have the form

$$t_2 = (\varepsilon_{xx})^2 \, g\left(|\varepsilon_{xx}|^2\right) \quad . \tag{12}$$

This gives us the phase information needed to assess coherence. The received harmonic amplitude from the crack at $\underset{\sim}{r}_n$ can be seen to be proportional to

$$\exp\{i[k_2|\underset{\sim}{r} - \underset{\sim}{r}_n| - 2\omega(t - t_n)]\} \left(\frac{X_n}{r_n}\right)^2 \exp[2i(kr_n - \omega t_n)] \quad . \tag{13}$$

Thus the phase ϕ has the form

$$\phi_n = 2k(|\underset{\sim}{r} - \underset{\sim}{r}_n| + r_n) + \psi \tag{14}$$

where ψ does not depend on n. Since the angular aperture is fairly small, one can expand to obtain

$$\phi_n \cong 2k\left((x - x_n) + x_n\right) + ky_n^2\left(\frac{1}{x - x_n} + \frac{1}{x_n}\right) + \psi \quad . \tag{15}$$

Thus, the harmonic contribution from M cracks is reduced below the coherent value by the magnitude of the factor F

$$F = \frac{1}{M} \sum_{n=1}^{M} \left(1 - \frac{y_n^2}{r^2}\right) \exp\{ikx\, y_n^2/[x_n(x - x_n)]\} \quad . \tag{16}$$

If the summation is less than or of the order of $M^{1/2}$ the superposition is incoherent. If the summation is of the order M ($F \sim 1$), the superposition is coherent. For the geometry of this experiment, the effective x value is about 90 mm. The wavelength of a 5 MHz surface wave in aluminum is approximately 580 μm. Thus, the argument of the exponential,

$$kx\left(\frac{y_n}{x}\right)^2 [(x_n/x)(1 - (x_n/x)]^{-1} \sim 2\pi \, \frac{(90)}{0.6} \left(\frac{2}{90}\right)^2 = 0.46 \quad . \tag{17}$$

The cosine of this representative value is 0.9. It is clear that, since the terms of the sum all have the same sign and are each of order unity, that $F \sim 1$, and the harmonic generation by each microcrack add essentially coherently.

Comparison to Results

It appears that the coherent generation model gives the best representation of measured harmonic amplitudes. The abrupt increase in A_2 early during the fatigue of the 4 μm α phase material is a consequence of a rapid initial burst of crack initiation. With continued fatigue A_2 falls as these cracks propagate beyond the grain of initiation. It is known that small asperities tend to hold noncrystallographic cracks in titanium partially open even at the maximum compressive load used during fatigue (16). Apparently, the opening is sufficient to prevent substantial harmonic generation. The continued increase in A_2 with fatigue in the 8 μm α phase material is due to the continued and accelerated initiation of new cracks. For both materials, approximately one hundred grain-sized cracks in the gauge section contributed to A_2 at the point of maximum harmonic amplitude. Coherent summation of these contributions would produce a received A_2 20 dB stronger than for incoherent summation, and it is likely that coherency is essential for detection of such a small number of microscopic cracks. Progressive increases in A_2 with fatigue has also been reported in aluminum alloys (15). It appears that this is not primarily due to initiation of new crystallographic cracks. Crack closure is more complete in aluminums than in titanium, apparently because aluminum is more ductile. As a consequence harmonics are generated in aluminum even by cracks many grain diameters in length.

Conclusions

The progressive increase in the anelasticity of the surface of Ti 6Al-4V alloy as surface microcracks develop during fatigue has been characterized by using an acoustic harmonic technique. It appears that in the small grain sized materials studied, acoustic harmonics are preferentially generated by grain sized surface cracks which are crystallographic. These close more fully and permit the development of the small normal loads over the large areas of the crack face which are required for generation of harmonics. A model of the harmonic generation process is proposed which sums the contribution of second harmonic signals produced by the individual grain-sized cracks developed by fatigue in the specimen gauge section. Agreement between the predicted and measured value is achieve by assuming that the individual contributions sum coherently. The received harmonic amplitude varies with fatigue as determined by the numbers of grain-sized cracks present. The amplitudes appear to be of sufficient magnitude to permit their use for estimating the average remaining fatigue lifetime of the titanium alloy in low cycle fatigue. In actual application of this technique, it is anticipated that larger amplitudes of the fundamental acoustic wave would be used to insonify the component to be inspected. This would further increase the ratio of the harmonic amplitude produced from microcracks to that developed from the dislocation motion which constitutes the background noise.

Acknowledgements

This research was supported by DARPA under Contract Number MDA 903-80-C-0641. The authors wish to thank Dr. M.R. James for residual stress measurements and discussion on the paper, and Mr. R. Govan for specimen preparation.

References

1. M.A. Breazeale and D.O. Thompson, Appl. Phys. Lett., 3 (1963) p. 77.

2. A.S. Gedroits and V.A. Krasil'nikov, Sov. Phys. - JETP, 16 (1963) p. 1122.

3. F.R. Rollins, Apply. Phys. Lett., 2 (1963) p. 147.

4. A. Hikata, B. Click and C. Elbaum, J. Appl. Phys., 36 (1965) p. 229.

5. F. Rischbieter, Acoustica, 16 (1965/66) p. 75.

6. F. Rischbieter, Acoustica, 18 (1967) p. 109.

7. C.R. Scorey, J. Appl. Phys., 41 (1970) p. 2535.

8. J.V. Yermilin, L.K. Zarembo, V.A. Krasil'nikov, Ye.D. Mezintsev, V.M. Prokhorov, and K.V. Khilkov, Phys. Met. Metallogr., 36 (1973), p. 174.

9. O. Buck, IEEE Trans. Sonics Ultrason., SU-23 (1976) p. 283.

10. A.V. Granato and K. Lucke, J. Appl. Phys., 27 (1956) p. 583.

11. J.C. Grosskrentz, Phys. Status Solidi B, 47 (1971) p. 11; 47, (1971) p. 359.

12. J.M. Richardson, Int. J. Engng. Sci, 17 (1979) p. 73.

13. O. Buck, W.L. Morris and J.M. Richardson, Appl. Phys. Lett., 33 (1978), p. 371.

14. W.L. Morris, R.V. Inman, and O. Buck, J. Appl. Phys., 50(11) (1979) p. 6737.

15. O. Buck, W.L. Morris and M.R. James, J. Non-Dest. Eval., 1 (1980), p. 3.

16. M.R. James and W.L. Morris, Unpublished research.

17. W.L. Morris, M.R. James and O. Buck, Met. Trans., 12A (1981), p. 57.

18. W.J. Pardee, J. Acoust. Soc. Am., (1) (1982), p. 71.

SOME INNOVATIVE TECHNIQUES FOR

NONDESTRUCTIVE EVALUATION OF MATERIALS

Robert E. Green, Jr.

Materials Sciences Division*
Defense Sciences Office
Defense Advanced Research Projects Agency
Arlington, VA 22209

Several innovative techniques for nondestructive evaluation of materials which have been developed at The Johns Hopkins University will be described in the present paper. The first of these is the rapid imaging of x-ray diffraction patterns using conventional and pulsed generators coupled with electro-optical detectors. These systems have been used to orient single crystals, to study lattice rotation accompanying plastic deformation, to measure the rate of grain boundary migration, to determine the physical state of exploding metals, to monitor the amorphous to crystalline phase transformation of rapidly solidified metals, to rapidly measure residual stress and to record topographic images of lattice defects in metal, semiconductors, and polymer crystals. Second, a set of experimental techniques will be described in which ultrasonic velocity and attenuation measurements give new insight into mechanical deformation processes during tensile elongation, compression, fatigue testing, and high-power insonation of materials. Finally, some recent experimental advances in acoustic emission detectors, involving laser beam interferometers, will be described which offer many advantages over conventional piezoelectric transducers and hold great promise for the future.

---

*Materials Science and Engineering Department
The Johns Hopkins University
Baltimore, MD 21218
(Permanent Address)

Rapid Display of X-ray Diffraction Images

The desirability of obtaining x-ray diffraction images with extremely short exposure times to study dynamic material alterations has long been realized. A review of the performance characteristics of electro-optical systems for rapid display of x-ray diffraction images has shown that a multiple stage image intensifier system coupled to an external fluorescent screen is the most sensitive and only truly instantaneous system (1). Figure 1 shows schematic drawings of two such devices which have been used for large format recording of Laue patterns as well as for small format recording of x-ray topographs.

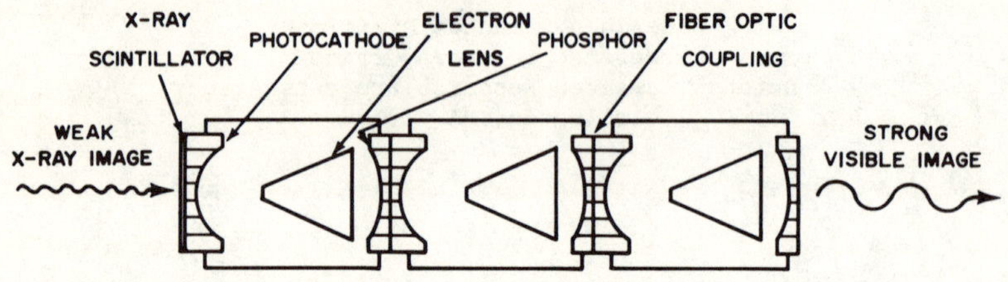

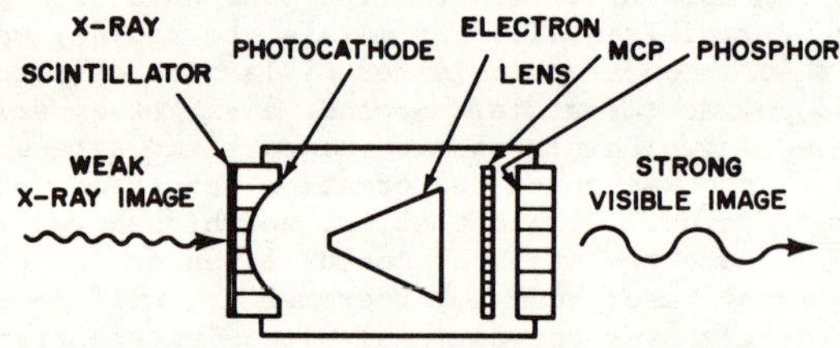

Figure 1. Schematic drawings of first generation (upper) and second generation (lower) x-ray image intensifier tubes for rapid display of x-ray diffraction images.

A first generation x-ray image intensifier tube has been used routinely for many years as a detector with the transmission Laue x-ray diffraction technique for determination of the orientation of single crystals and large grains in polycrystalline aggregates. Experiments have also been conducted in which crystallographic orientation changes due to lattice rotation were continuously monitored during uniaxial tensile deformation, (2,3) in which grain boundary migration measurements were continuously made at elevated temperatures, (4) in which the amorphous to crystalline phase transformation of rapidly solidified metals was continuously monitored, (5) and in which rapid residual stress measurements were made (6).

In 1973, a review paper (7) was published which traced the development of generation and detection systems for flash x-ray diffraction and summarized the state-of-the-art of such systems. A comparative evaluation was presented of flash x-ray diffraction systems using generators which rely on increased electron beam current and those which rely on higher voltage. Comparison was also made between detection systems incorporating film recording, scintillators fiber-optically coupled to photomultiplier tubes, and image-intensifier systems both lens and fiber-optically coupled to fluorescent screens. A more recent review was presented in 1977, which was directly concerned with the applications of flash x-ray diffraction systems to materials testing (8). The x-ray diffraction patterns obtained using such flash x-ray diffraction systems have yielded detailed information about microstructural alterations caused by explosions, heat pulses, and shock waves.

Electro-optical systems have also been used as detectors for rapid recording of x-ray topographic images. A paper (9) presented in 1977 gave a comprehensive review of all electro-optical systems used up to that time for direct viewing of x-ray topographic images. Consideration was given to both direct conversion x-ray sensitive vidicon systems and to indirect conversion systems which use fluorescent screens to convert the x-ray image into a visible one. In the direct method an x-ray sensitive vidicon television camera directly converts the x-ray topographic image into an electronic charge pattern on a lead oxide or silicon diode array target; this charge pattern is read out by a scanning electron beam and displayed as a visible image on a television monitor.

In the indirect method the x-ray topographic image is converted into a visible light image by a fluorescent screen. This visible light image is then optically coupled either by a lens or a fiber-optic faceplate to the input photocathode of a low-light-level electro-optical device. Depending on the type of electro-optical device used, the output image may either be viewed directly or viewed on a television monitor. This method has been reported to yield a topographic image of approximately 10 μm resolution which is limited by the grain size and thickness of the fluorescent screen. This high resolution requires a fine grained thin fluorescent screen which has low conversion efficiency and hence necessitates use of a high intensity rotating anode x-ray generator or synchrotron radiation. By increasing the thickness or absorption coefficient in the correct wavelength regime of the fluorescent screen, the conversion efficiency can be increased to the extent that conventional low intensity x-ray generators can be used.

Ultrasonic Velocity and Attenuation Measurements

The use of ultrasonic waves as nondestructive probes has as a prerequisite the careful documentation of the propagational characteristics of the ultrasonic waves themselves (10). Since in nondestructive evaluation applications it is not desirable for the ultrasonic waves to alter the material through which they pass, it is necessary to work with very low amplitude waves, which normally are regarded to obey linear elasticity theory. In general, three different linear elastic waves may propagate along any given direction in an anisotropic material. These three waves are usually not pure modes since each wave generally has particle displacement components both parallel and perpendicular to the wave normal. However, one of these components is usually much larger than the other; the wave with a large parallel component is called quasi-longitudinal, while the waves with a large perpendicular component are called quasi-shear. In the event that sufficient symmetry prevails such that the direction of propagation is elastically isotropic or if the material is elastically isotropic in general, then all modes become pure modes, i.e. the particle displacements are either parallel or perpendicular to the wave normal, and the two quasi-shear modes degenerate into one pure shear mode. Also of great importance to elastic wave propagation in anisotropic materials is the flow of energy contained in the wave. For linear elastic wave propagation in anisotropic materials, the direction of the flow of energy per unit time per unit area, the energy-flux vector does not, in general, coincide with the wave normal as it does in the isotropic case, i.e. the ultrasonic beam exhibits refraction.

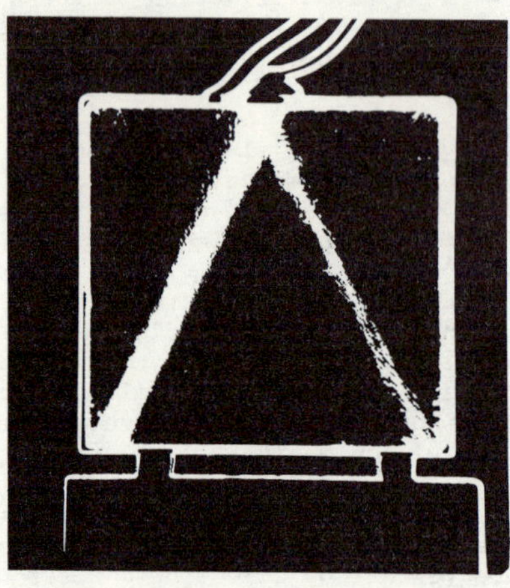

Fig. 2. Schlieren photograph of ultrasound field associated with simultaneous generation of quasi-longitudinal and quasi-shear waves in a quartz crystal from a single x-cut piezoelectric transducer.

Figure 2 serves to illustrate several aspects of ultrasonic wave propagation in anisotropic materials. This figure is one of a set of four photographs originally published by Merkulov and Yakovlev (11) who used the optical schlieren technique to visualize the sound field due to ultrasonic waves propagating in optically transparent quartz crystals. In Fig. 2 a single 25 MHz x-cut quartz piezoelectric transducer attached to a y face of a single crystal quartz block is shown to simultaneously generate a quasi-longitudinal and quasi-shear wave, both of which exhibit energy-flux deviation from the wave normal. In general, a piezoelectric transducer generates both longitudinal and shear waves, but depending on the particular piezoelectric properties of the specific transducer one type of mode will be dominant. It should be clear, that since most real structural materials exhibit some degree of anisotropy, great care must be exercised in locating flaws in these materials using ultrasonic techniques. For example, it would be erroneous to assume that a crack intersecting the quasi-longitudinal wave, shown on the left side of Fig. 2, such that a measurable amount of energy was reflected back to the transducer, was directly beneath the transducer as it would be in the case of an isotropic material.

Here is a good place to call attention to the fact that, although many authors and investigators draw a close analogy between electromagnetic (light) wave propagation and elastic (sound) wave propagation in solid materials, great caution should be exercised in doing so. The behavior of anisotropic materials with respect to propagation of elastic waves is much more complicated than the propagation of electromagnetic waves, since the material constitutive equations required to properly describe elastic waves are of higher order tensor quantities than those required to describe electromagnetic waves.(12)

There are various mechanisms by which energy can be lost from ultrasonic waves propagating through real materials, and measurement of this ultrasonic attenuation can complement ultrasonic velocity measurements in yielding valuable information about the mechanical properties of the material. Once proper precautions are taken to either eliminate or control geometrical effects, ultrasonic attenuation measurements serve as a very sensitive indicator of internal loss mechanisms caused by microstructures and microstructural alterations in metallic materials.(13) This sensitivity derives from the ability of ultrasonic waves to interact with volume defects such as cracks, microcracks, included particles, and precipitates; surface defects such as grain boundaries, interphase boundaries, and magnetic domain walls; and dislocation line defects. Moreover, ultrasonic waves are sensitive to the interaction of point defects such as impurity atoms and vacancies with dislocations.

Figure 3 demonstrates the ability of ultrasonic attenuation measurements to detect extremely small microstructural alterations during fatigue testing and hence give early warning of fatigue induced failure. This figure was taken from the work of Joshi and Green (14) who used ultrasonic attenuation measurements as a continuous monitor of fatigue damage in cold-rolled 1015 polycrystalline steel specimens which were cycled in reverse bending as cantilever beams to fracture at 30 Hz. Note that the ultrasonic attenuation change indicated microstructural alterations, probably microcrack formation, prior to detection of an additional ultrasonic pulse on the A-scan display as is used in conventional non-destructive ultrasonic evaluation to detect crack formation.

Although x-ray diffraction techniques have historically been the nondestructive method most often used in actual practice to measure residual stress, they are not optimally suited for field applications partially because the necessary equipment is heavy and bulky, and, perhaps more importantly, they suffer from the fact that they only serve to determine the state of stress in a surface layer of a material, while in many practical cases a knowledge of the bulk stresses is desired. Ultrasonic residual stress measurements have been made far less frequently, (10) probably because of lack of familiarity with nonlinear elasticity theory and the experimental techniques necessary to obtain the required ultrasonic velocity measurement accuracy. Most ultrasonic measurements of residual stress have been based on stress-induced acoustical birefringence of a homogeneous isotropic solid. When shear waves are propagated through a homogeneous isotropic solid at right angles to the direction of applied stress, the shear wave with particle displacement parallel to the direction of applied stress propagates with a faster velocity than the shear wave with particle displacement perpendicular to the direction of the applied stress. Other ultrasonic techniques such as harmonic generation, temperature dependence of ultrasonic velocity, or ultrasonic dispersion have been proposed for residual stress determinations.

The major difficulty associated with reliable ultrasonic residual stress measurement is that the change in elastic wave velocities in solid materials due to residual stress is small and other factors which can cause greater velocity changes may mask residual stress effects. The prime factor in this regard is the fact that all real structural materials are initially anisotropic and not isotropic. The solidification and/or forming and heating

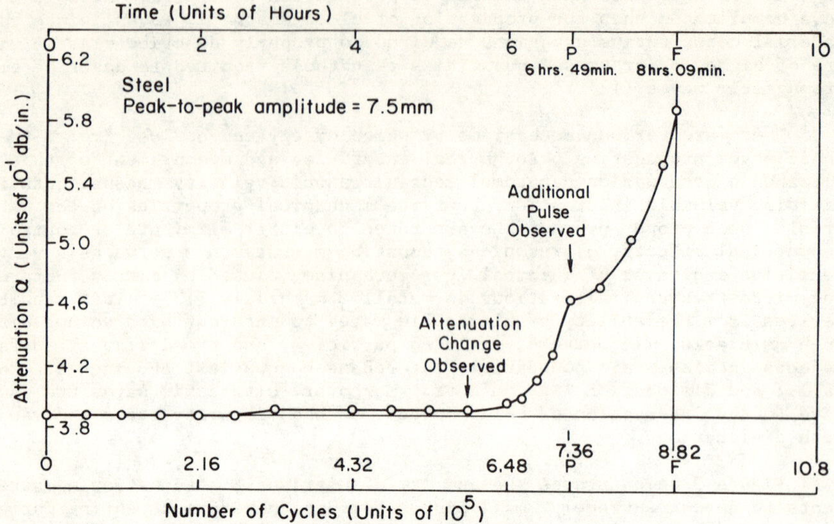

Figure 3. Longitudinal ultrasonic wave attenuation versus number of fatigue cycles for cold-rolled 1015 polycrystalline steel specimen.

processes associated with metal preparation does not permit a random distribution of crystallographic orientations among the grains of the polycrystalline aggregate and often does not even permit a uniformity in grain size. Therefore, real polycrystalline metals possess a "texture" and this texture strongly influences the mechanical properties of the metal including ultrasonic wave propagation. Thus, the ultrasonic residual stress measurement problem becomes one of measuring changes in anisotropy before and after stressing rather than the commonly treated theoretical problem of stress-induced anisotropy of an originally isotropic solid.

Acoustic Emission Source Characterization

Acoustic emission is the phenomenon of transient elastic wave generation due to a rapid release of strain energy caused by a structural or microstructural alteration in a solid material. In order to positively identify the source of an acoustic emission signal and hence be able to make a definitive statement as to whether or not the material alteration causing the acoustic emission signal is harmful to the integrity of the engineering structure, it is necessary to be able to determine the unmodified waveform and frequency spectrum of the signal itself. Thus, a detection system must be used which is capable of sensing surface displacements due to both internal and surface acoustic emission sources which does not modify the signal because of its own limitations. Although they have been used in the past, and are still the detectors most often used in current practice, commercial piezoelectric transducers are completely unsuited for acoustic emission waveform measurements. In addition to possessing their own amplitude, frequency, and directional response, they "ring" at their own resonance frequency and it is impossible to distinguish between the amplitude (voltage output) excursions caused by this "ringing" and the amplitude variations actually characteristic of the acoustic emission source. Also, since the commercial piezoelectric transducers must be coupled to the workpiece, the waves being measured are physically disturbed.

In order to optimally solve the detector problem several laser beam optical interferometric systems for detection of acoustic emission signals have been developed. (15) These optical detectors offer many important advantages over commercial piezoelectric transducers for monitoring acoustic emission events. Among these advantages are: direct contact with the test specimen is unnecessary; no acoustical impedance matching couplant is required; the waveform and frequency spectrum of the acoustic emission event is not modified by the optical probe; optical probes have inherent broad flat frequency responses; they can probe internally in transparent media; they can be used to make measurements on extremely hot and extremely cold materials and in other environments hostile to piezoelectric transducers; since the focused laser beam diameters are typically only a few hundredths of a millimeter, they can probe very close to a slip band, mechanical twin, included particle, crack, or other material defect.

Optical probes have already been used to measure acoustic emission waveforms due to mechanical twinning of cadmium, indium, tin, and zinc; cracking of glass by both thermal and mechanical means; stress-corrosion cracking in E340 steel and 7039 aluminum; allotropic transformation in iron; and plastic deformation and fracture of a number of stainless steel, aluminum, and titanium alloy specimens.

Figure 4 compares 200 μsec waveforms from fracture of a glass capillary against an aluminum test block as (a) with a non-contact laser interferometer and (b) with a direct contact conventional commercial piezoelectric transducer. The portion of the laser interferometrically detected signal to the left of the vertical dashed line agrees within experimental accuracy with the theoretical predictions of the displacement associated with the source.

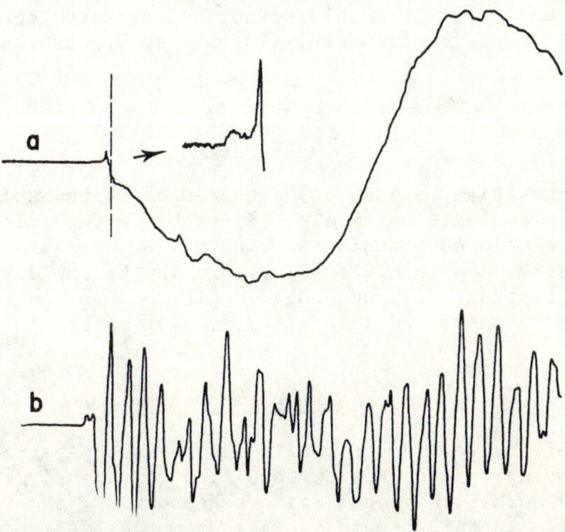

Fig. 4. Comparison of acoustic emission waveforms from fracture of a glass capillary as detected with (a) laser interferometer and (b) conventional commercial piezoelectric transducer, 200 μsec time records.

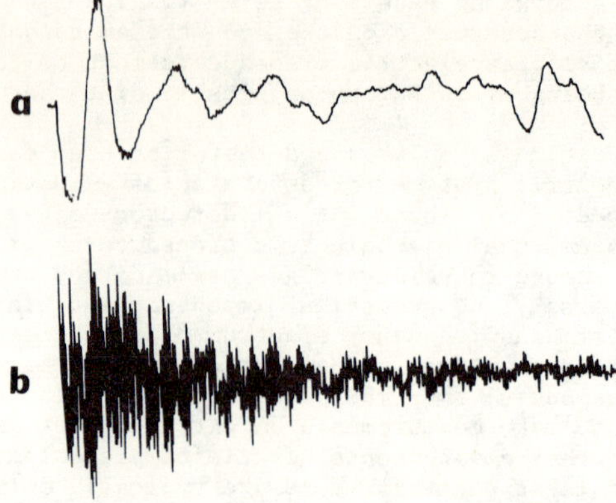

Fig. 5. Longer time records, 2 msec, of same two waveforms as shown in Fig. 4.

The arrow indicates an enlarged view of this portion of the waveform. Even longer time records, 2 msec, of the same two waveforms are reproduced in Fig. 5. As can be seen in the figures, the commercial piezoelectric transducer introduces a multitude of oscillations into the detected waveform and does not faithfully reproduce the waveform due to surface displacements characteristic of the acoustic emission source.

Although the literature is saturated with claims of acoustic emission source identification due to a variety of microstructural alterations in real materials, all of these measurements have been made on large specimens with commercial piezoelectric transducers either located on the specimens at locations remote from the source, or, even worse, located on specimen grips or other such supports. It should be obvious from the considerations presented here, and in more detail elsewhere, (15) why such measurements must be viewed with skepticism.

Conclusion

Several innovative techniques have been described which can be used to nondestructively evaluate materials. Space has not permitted description of other recently developed techniques which may be of equal or even greater importance. These, and future developments in the rapidly expanding field of nondestructive testing, will undoubtedly play an ever increasing role in the processing, fabrication, and life assessment of materials.

Acknowledgements

Portions of the work described in the present paper have been sponsored by Aberdeen Proving Ground, Air Force Office of Scientific Research, Army Research Office Durham, National Science Foundation, and the Naval Sea Systems Command. In this regard the author would like to especially thank Dr. George Mayer (AROD), Dr. Robert Reynik (NSF), and Dr. Hans Vanderveldt (NAVSEA). Thanks are also due Mrs. Sydney Green for assistance with preparation of the camera-ready manuscript. Alma Comer deserves special acknowledgement for her professionalism and care in typing the manuscript.

References

1. R. E. Green, Jr., in Advances in X-ray Analysis, edited by C. S. Barrett, John B. Newkirk, and C. Ruud (Plenum, New York, 1971), Vol. 14, pp. 311-337.

2. K. Reifsnider and R. E. Green, Jr., Trans. Met. Soc. AIME 246, 1615-1619 (1969).

3. N. R. Joshi and R. E. Green, Jr., J. Mater. Sci. 15, 729-738 (1980).

4. R. E. Green, Jr., in Advances in X-ray Analysis, edited by K. F. J. Heinrich, C. S. Barrett, J. B. Newkirk, and C. O. Ruud (Plenum, New York, 1972), Vol. 15, pp. 435-445.

5. C. Beczak, R. B. Pond, Sr., and R. E. Green, Jr., The Johns Hopkins University (unpublished work).

6. D. A. Steffen and C. O. Ruud, in Advances in X-ray Analysis, edited by C. S. Barrett, D. E. Leyden, J. B. Newkirk and C. O. Ruud (Plenum, New York, 1978), Vol. 21, pp. 309-315.

7. J. A. Dantzig and R. E. Green, Jr., in Advances in X-ray Analysis, edited by L. S. Birks, C. S. Barrett, John B. Newkirk, and C. O. Ruud (Plenum, New York, 1973), Vol. 16, pp. 229-241.

8. R. E. Green, Jr., in Proceedings of the Flash Radiography Symposium, edited by L. E. Bryant, Jr. (The American Society for Nondestructive Testing, Columbus, Ohio, 1977), pp. 151-164.

9. R. E. Green, Jr., in Advances in X-ray Analysis, edited by H. F. McMurdie, C. S. Barrett, J. B. Newkirk, and C. O. Ruud (Plenum, New York, 1977), Vol. 20, pp. 221-235.

10. R. E. Green, Jr., Ultrasonic Investigation of Mechanical Properties, Vol. 3, Treatise on Materials Science and Technology (Academic, New York, 1973).

11. L. G. Merkulov and L. A. Yakovlev, Akust. Zh. 8, 99-106 (1962) [Sov. Phys.-Acoust. 8, 72-77 (1962)].

12. E. G. Henneke, II, and R. E. Green, Jr., J. Acoust. Soc. Am. 45, 1367-1373 (1969).

13. R. E. Green, Jr., in Nondestructive Evaluation: Microstructural Characterization and Reliability Strategies, edited by O. Buck and S. M. Wolf (The Metallurgical Society of AIME, Warrendale, Pennsylvania, 1981), pp. 115-132.

14. N. R. Joshi and R. E. Green, Jr., Eng. Fract. Mech. 4, 577-583 (1972).

15. R. E. Green, Jr., in Mechanics of Nondestructive Testing edited by W. W. Stinchcomb (Plenum, New York, 1980), pp. 55-76.

NONDESTRUCTIVE METHOD FOR DETERMINING FATIGUE AND STRESS-CORROSION DAMAGE

I. R. Kramer

Technical Advisor
David W. Taylor Naval Ship R&D Center
Annapolis, Maryland 21402
and
Research Professor
University of Maryland
College Park, Maryland 20740

and

S. Weissmann
Professor, Department of Mechanics and Material Science
College of Engineering
Piscataway, New Jersey 08854

Abstract

The distribution of dislocations in the surface layer and the interior of fatigued and stress-corrosion damaged specimen was determined by X-ray line broadening techniques. The data indicate that in both cases the dislocation density in the region of the surface is higher than that in the interior initially. With increasing damage the dislocation density in the interior increases rapidly and finally becomes equal to that of the surface. A crack or fracture occurs when this condition is fulfilled. The ratio of the dislocation density in the surface and the interior can provide a measure of the damage caused by fatigue and stress corrosion. Measurements of fatigue and stress-corrosion damage by the X-ray technique agreed closely with the measured damage.

Introduction

It is obviously very desirable to be able to determine by nondestructive methods the amount of fatigue and stress-corrosion damage. The ability to achieve these goals has not been attained to any considerable degree in spite of the considerable efforts devoted to the subject. In general, most of these investigations have been directed toward investigations concerned with microstructural or defect changes that occur in the bulk of the material, or such changes that occur at the surface, per se. One of the important points to be made in this paper is that fatigue and stress-corrosion damage may be determined when both the surface layer together with the bulk material are taken into consideration. Either taken separately may lead to erroneous conclusions.

In a series of investigations starting in 1961, it was shown that in the region of the surface to a depth in the range of 50 to 100 μm, the work-hardening was greater than that in the bulk of the material (1,2). This layer was also reported to form in fatigued specimens (3,4). It was further shown that in stress-corrosion cracking, those environments that caused damage also increased the work hardening of the surface region (5). These investigations led to the concept that the surface influences to a considerable degree, and in some cases controls, the dislocation generation in the bulk especially when the fatigue and stress-corrosion experiments are conducted below the macroscopic yield stress.

In a series of studies on aluminum 2024-T6, titanium (6Al/4V), and a 4130 steel the surface layer stress was measured as a function of the number of fatigue cycles under high-cycle conditions. The surface layer stress was defined as the additional stress required to produce a given amount of plastic deformation because of the extra work hardening in the surface layer. For these three classes of materials (fcc, hcp, bcc), a propagating crack formed whenever the surface layer stress attained a critical value for each material. This value was independent of the stress amplitude, the environment, and prior cyclic history. These observations are in agreement with the concept that crack formation at these low applied stresses is associated with an accumulation of dislocation of like sign (excess dislocations) in the surface layer (4,6). According to this concept, to form a crack, the surface layer acts as a barrier to a piled-up array or a local accumulation of dislocations of like sign and, at a sufficiently high applied stress and dislocation density, the stress field ahead of the "pile-up" exceeds the local fracture strength of the material. At low barrier strengths, plastic flow ahead of the "pile-up" occurs to allow stress relaxation. As will be discussed later in more detail, the formation of a surface layer containing a high density of dislocation is a necessary but not a sufficient condition for crack formation. It will be shown in those cases investigated that a crack forms when the dislocation density in the interior becomes essentially equal to that of the surface layer, and the dislocation density is essentially uniform throughout the cross section of the specimen.

X-Ray Diffraction Study of Surface Layer

X-ray diffraction analysis can provide a reliable method for the measurement of the changes in the dislocation density as a function of plastic deformation. The techniques are highly sensitive to the changes in the crystal structure and may be employed for a nondestructive evaluation. Both the double crystal diffractometer and the powder X-ray techniques have been employed in these investigations to determine the changes in the dislocation density-depth profile of specimens subjected to tensile deformation and fatigue. The dislocation density, ρ, may be related to β, the widths of the X-ray line, or spots by (7):

$$\rho = \beta/bT, \tag{1}$$

where T is the subgrain size and b is the Burger's vector or

$$\rho = \beta^2/9b^2. \tag{2}$$

Equations (1) and (2) give the lower and upper bounds of the dislocation density, respectively. The dislocation density may be obtained from an analysis of the line profile (8) and the method of integral breadths (9). According to the latter

$$\beta \cos \theta = \frac{K\lambda}{L} + 4e \sin \theta, \tag{3}$$

where $e = \Delta d/d$ and d is the interplaner spacing. The first term on the right is related to the particle size while the second term is related to the microstrain $\langle \varepsilon^2 \rangle^{\frac{1}{2}}$. The dislocation density associated with the particle size and strain is, respectively

$$\rho_p = \frac{1}{L^2}, \tag{4}$$

$$\rho_\varepsilon = 12\langle \varepsilon^2 \rangle^{\frac{1}{2}}/b^2 \tag{5}$$

where $\langle \varepsilon^2 \rangle^{\frac{1}{2}} = e/1.25$ is the root mean square microstrain.

Dislocation-Depth Profile of Unidirectional Stressed Crystals

A somewhat typical X-ray line profile obtained for unidirectional strained metals is shown in Figure 1a for a silicon crystal that was strained 10% at 650°C. In this example, a double crystal diffractometer was employed. After the deformation, 100 μm was removed and it is apparent that the line sharpened considerably (Figure 1b). Figure 2 shows the dislocation density-depth profile (ρ-x) for silicon, a material with a low stacking fault energy; aluminum, a high stacking fault energy material; and gold, which exhibits little or no propensity for the formation of a surface oxide. To prevent possible relaxation effects in aluminum, the straining and X-ray measurements were carried out at -5°C. The ρ values given in Figure 2 and in the other portions of this paper were calculated from Equation (2). For all three metals, there was a decline by factors of 3 to 6 from the high surface layer density, ρ_s, to a constant value, ρ_i, at about 100 to 150 μm in the interior. These results, together with the published data, demonstrate that the work hardening even in unidirectional straining is not uniform throughout the cross section.

Dislocation Density-Depth Profile of Fatigued Specimens

Figure 3 shows data obtained by analysis of the surface layer, using shallow-penetrating copper radiation, ~7μm, after cycling specimens of aluminum 2024-T3 for various fractions of the fatigue life. The results are plotted for specimens with two average grain sizes differing by about 25% in grain diameter, and for tests carried out at two different stress amplitudes. In agreement with Taira, et al (10, 11), the change in excess dislocation density during the life could be described by a three-stage sequence for all the tests. Rapid increases in the defect concentration occurred early (Stage I) and late (Stage III) in the life. A markedly decreased slope was obtained for the long duration of the second stage comprising the period from 0.15 to 0.95 N_f. All the data fell within the experimental error band, even for the alternate grain size stock, if a uniform shift factor related to the ratio of grain diameters was applied (12).

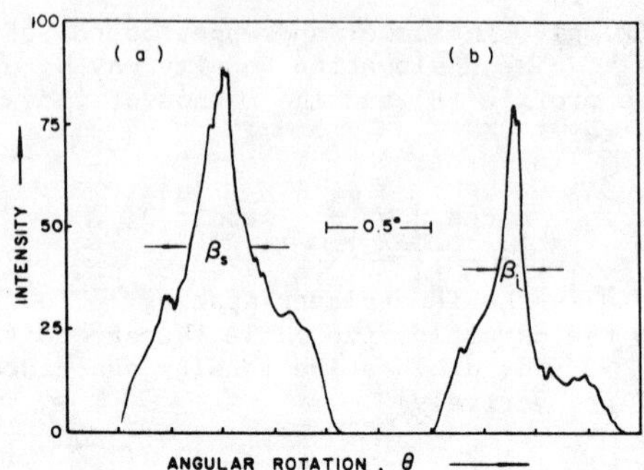

Figure 1 - X-ray rocking curve profiles of tensile deformed silicon single crystal. ε_p = 10% 650°C, tensile axis [110], (112) reflection, Cu Kα. (a) Original surface, (b) after removal of a 100 μm surface layer.

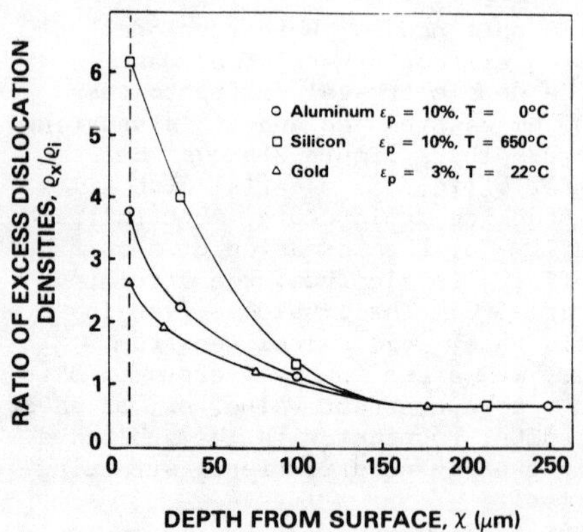

Figure 2 - Distribution of excess dislocations with depth from the crystal surface. Tensile axis and surface orientations: Al 100 and (100); Si 110 and (112); Au, 123 and (311)

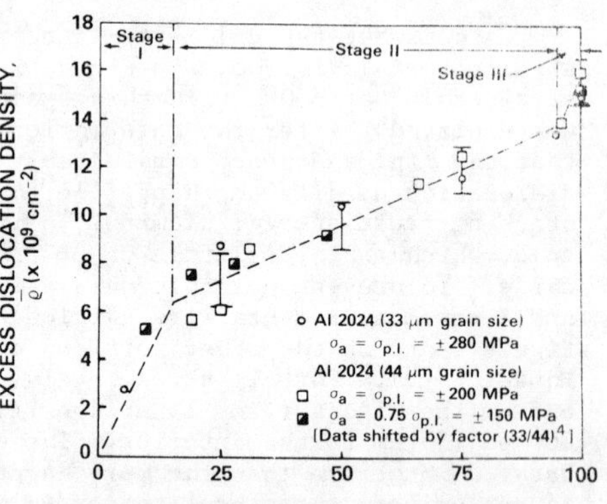

Figure 3 - Excess dislocation density in the surface for fatigued Al 2024-T3 specimens with G.S 4 and 33 mm, tested at various stress amplitudes. Cu Kα radiation.

The data presented in Figure 4a were obtained from the rocking curve analysis of aluminum 2024-T3 specimens cycled for various fatigue life fractions at ±200 MPa corresponding to the proportional limit. Analogous to the deformation characteristic of monotonically and cyclically deformed single crystals, the ρ-x profile revealed a higher excess dislocation density in the surface layer than in the bulk. Up to about 0.15% of the fatigue life (N = 21,000), the ρ-x profile was similar to that observed after single tension. Observations made after 5% of the fatigue life showed that the dislocation density in the bulk also increased, and a trough appeared on the ρ-x profile in the sub-surface region. With further cycling, the excess dislocation density continued to increase in the surface layer and in the bulk.

The data in Figure 5 were obtained for specimens of aluminum 2024-T4 fatigued in 3.5% NaCl in axial tension-tension. Except for the absence of a minimum, the curves are similar to those obtained for the axial tension-compression cases shown in Figure 4. At the lower damage level of 31% and 67% the β values to a depth of about 50 μm are greater than those in the bulk material. As the fatigue damage increased, the β values of the interior increased more rapidly than those of the surface, and again failure occurred when the two were essentially equal.

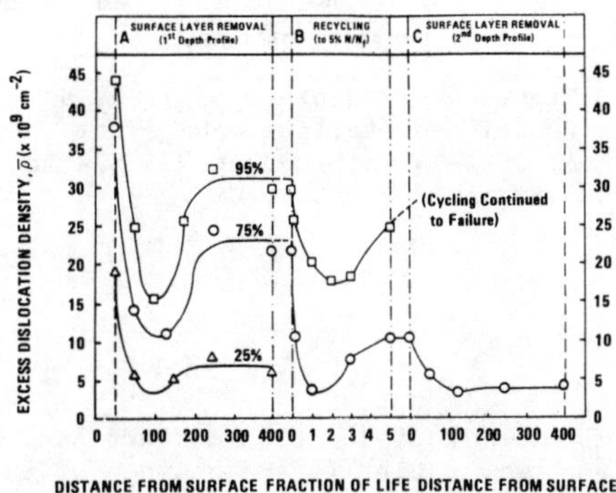

Figure 4 - (A) Excess dislocation density-depth profile for Al 2024-T3 specimens fatigued various percentages of life. (B) Dislocation density of surface after removal of surface layer and cycling. (C) Dislocation density - depth profile after fatiguing, N/N_f = 0.75, removal of surface and cycling N/N_f = 0.05.

Instability of Defect Structure in the Interior

From Figure 4 it is clear that ρ_s and ρ_i continue to increase with the amount of fatigue damage. The observation that ρ_i increases in the bulk appears to imply that permanent damage has occurred throughout the specimen, and yet it is known that the fatigue life of metals is completely recovered when the surface layer is removed periodically during the cycling process. The answer to this apparent conflict is that the defect structure in the interior is unstable without the presence of the surface layer. The instability of the interior defect structure is demonstrated in Figure 4b by specimens that had been cycled 75% and 95% of their life and after removal of 400 μm were cycled again at the same stress (200 MPa). The dislocation density determined from the rocking curves employing copper radiation declined very rapidly during the initial recycling and, after about 200 cycles, reached a minimum that was approximately the same as that of the original specimen. After 1% to 2% of the life, β increased again. When the recycling was continued to 5% of the fatigue life, the ρ-x profile over the entire cross section was the same as that of the virgin specimen fatigued 5% of the life.

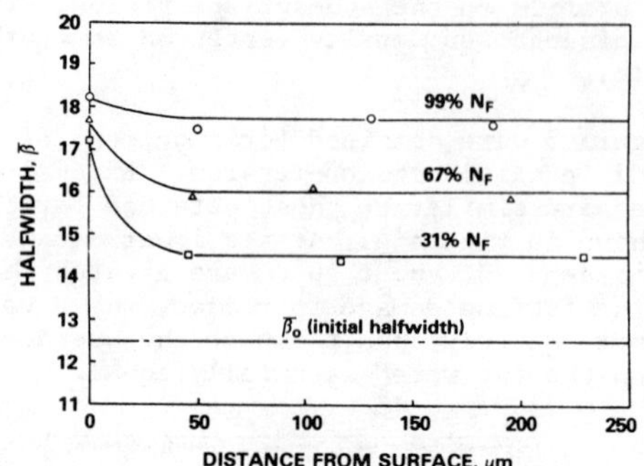

Figure 5 - β depth profile as a function of fatigue damage for Al 2024-T4 fatigued at 276 MPa in 3.5% NaCl. R = 0.1, Cr Kα

The reversion of the bulk dislocation content back to the virgin state when fatigued specimens were cycled in the absence of the work-hardened surface layer explains the extension of the fatigue life by periodic removal of the surface. The observations relative to this instability imply that there is a strong reaction of the dislocation in the interior to the surface layer. At the stress amplitude employed, it is apparent that the dislocation density in the bulk would not have increased during the cycling without the presence of the surface layer. In this sense, the increase in the bulk dislocation density is controlled entirely by the surface layer at least for high-cycle fatigue processes.

The instability observation has important implications relative to the influence of the environment on the work-hardening characteristics. The instability observations lead to the concept that the work-hardening of the interior is dependent on the surface layer. Accordingly, an environment that influences the surface layer will also affect the work-hardening of the bulk. For this to occur, the stress fields associated with the dislocation array in the surface layer must act over long distances. Conceiveably, if the dislocations are arranged in a form similar to an inverse "pile-up," the stress may be transmitted over long distances. According to this model, the effective stress acting on the dislocations in the interior has, in addition to the applied stress, a stress component due to the dislocation array of the surface layer. The proposal that the stress field associated with the surface layer can act over a long distance to influence the plastic deformation of the bulk material finds support in the investigations of Baranov, et al (13). Whereas it was previously shown that Stages I, II, and III were strongly influenced by removing the surface electrochemically during the deformation (14), these investigators reported that the surface removal also had a remarkable effect on the ductility of tungsten crystals. When no surface removal was involved, these single crystals normally failed in a brittle manner along (001) slip planes with a reduction in an area of about 12%. With surface removal, these crystals fractured with the formation of a neck and a reduction of an area of 83%. Apparently, since necking occurs when $d\sigma/d\varepsilon = \sigma$, the removal of the surface decreased the work-hardening of the entire specimen and local instability occurred at a stress much lower than the cleavage stress.

Dislocation Distribution in Surface and Bulk in Fatigue and Stress Corrosion

A very interesting and important aspect of fatigue failures as well as other subcritical crack formation such as stress-corrosion cracking and hydrogen embrittlement is that the cracks form at a relatively low stress level compared to the fracture stress of the material. It has been proposed (4,6) that such a crack can be formed by the pile-up of dislocation against or within the surface layer. According to this proposal, a crack would form when the surface layer became sufficiently strong to support a critical size pile-up. When the local stress associated with this accumulation of dislocations of like sign exceeds the fracture stress, a crack will form in the affected region. It should be noted that in Figure 3, and from Taira's results (10,11), fatigue failure occurred when the excess dislocation density attained a critical value, β^* or ρ^*, that was independent of the applied stress amplitude. It was also reported that propagating cracks formed when the surface layers reached a critical strength (6). The two cases are equivalent since $\sigma_s = \alpha G b \rho^{\frac{1}{2}}$. In this case, fracture occurred at the critical strength value independent of the environment, stress amplitude, and prior cyclic history. A necessary part of the surface layer fracture model is that a critical barrier strength is required to prevent relaxation of the stress fields. Fracture does not occur prior to the attainment of this critical strength, because the stress fields from the excess dislocations would be relaxed by plastic deformation. In fatigue processes where the direction of the dislocation motion changes with the direction of the applied stress, it is possible that stress relaxation can also occur.

It appears that in stress corrosion, the mechanism for damage may be similar to that occuring in fatigue. Figure 6 gives tha half width, β, as a function of distance from the surface for a 304 stainless steel stressed at 55% of the yield strength for various fractions of the cracking time, t/t_c, in 42% $MgCl_2$, where tc is the time to form a crack. The experimental data (15) show that a surface layer with a negative dislocation gradient is formed first. At $t_c = 0.1$, the β values in the interior are essentially the same as those of the unstressed material. With increasing exposure time in the $MgCl_2$ solution, the density of dislocation on the interior and the surface layer increases continuously. However, similar to the findings in fatigue, the increase in the β values in the interior is much more rapid than those at the surface. Again, similar to fatigue, cracking occurs when the values in the surface are essentially equal to those in the interior. In 1971, Kamachi and coworkers (16) showed (Figure 7) that the β values measured at the surface of 304 stainless steel specimen exposed to 42% $MgCl_2$ at various stresses increased continuously during the exposure time. The β values for specimens exposed to the same temperature and stress in an inert environment (parafin oil) increased relatively small amounts initially and after about 2 hours remained constant. It is apparent from these results that the environment caused a rapid increase in the generation of dislocations in the surface layer and in the interior as a consequence of the production of dislocations in the surface layer. Similar to the process proposed previously for fatigue (4) and stress-corrosion cracking (17,18), it appears that the stress field from dislocations in the region of the surface interacts with sources in the interior to cause a general increase in the dislocation density of the specimen as a whole.

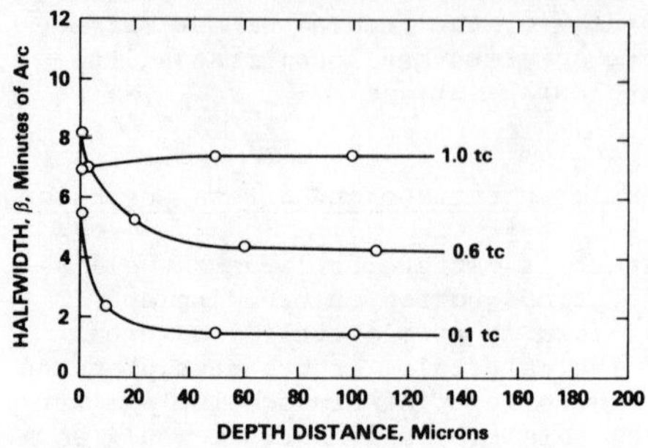

Figure 6 - β-x profile for 304 stainless steel stressed at 55% of the yield strength in $MgCl_2$ for various times

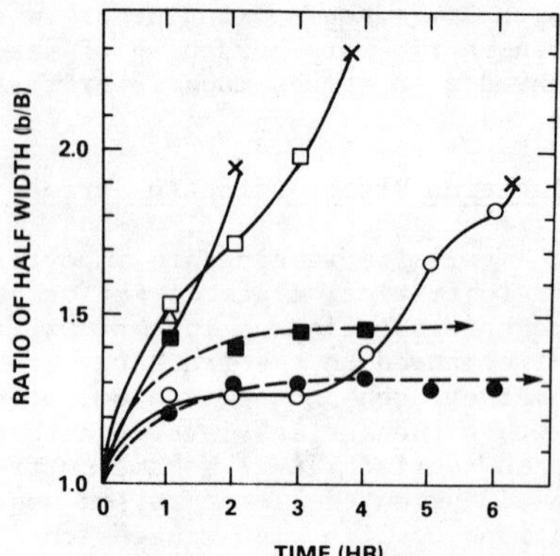

Figure 7 - β-time values measured at the surface for 304 stainless steel stressed in $MgCl_2$

Determination of Fatigue and Stress-Corrosion Damage

Of interest in Figure 3 is the observation that all the points fall on a single curve independent of the stress amplitude. Most importantly, a propagating crack apparently is initiated, and failure occurs when a critical value of the excess dislocation density, β^*, is attained. It is evident that if ρ^* can be determined, and the relationship between fatigue damage and the excess dislocation density is established, an assessment of the fatigue damage can be made. It may be seen from Figure 3, as well as from the data of Taira, et al (10,11), that the slope of Stage II is too low to be used for accurate and practical determination of fatigue damage. An examination of the depth profiles shown in Figure 4 reveals that the buildup of excess dislocation density at the surface layer occurs much earlier in the life than in the bulk. However, inspection of the plateau values, established at about 250 μm, discloses that the average excess dislocation density in the bulk increases steadily throughout the life. It follows that the fatigue damage can be determined, provided the X-ray beam penetrates sufficiently far into the materials to sample both the surface layer and bulk. Experimentally, for aluminum this can be achieved by application of X-ray radiation with a short wavelength, such as molybdenum Kα radiation. Figure 8 exhibits a plot which compares the average excess dislocation densities obtained from measurements with copper and molybdenum radiation. It is seen that by applying molybdenum radiation, a sharp incline of the slope is obtained up to the critical value, ρ^*, and enables a determination of the accumulated damage unequivovally. By contrast, application of copper radiation, which samples only the defect structure of the quickly saturated surface layer, cannot accomplish this task, owing to the shallow slope during the intermediate life fractions. It is important to note that at $N/N_f = 1$, the ρ^*, values measured with molybdenum and copper radiation are equal. Since the former penetrates about 350 μm and the latter about 7 μm, the observation indicates that the dislocation density is uniform throughout the cross section of the specimen. It is also possible to estimate β^* from surface measurements. From Figure 3, when plotted in terms of β instead of ρ, the slope in the region from 25% to 95% damage is very low. Not too much error is made if the value for β is selected from these surface measurements, provided they are obtained in the region where the β values are increasing very slowly with the number of cycles. Of course, if a calibration curve such as in Figure 8 is available, then the damage may be read directly from the curve. In practice, it is unlikely that such a curve will be available. Yet another procedure may be used to determine whether to remove a part from service. Without actually knowing β^*, the part may be removed whenever the β_s/β_i value exceeds some prechosen value, for example, 0.9.

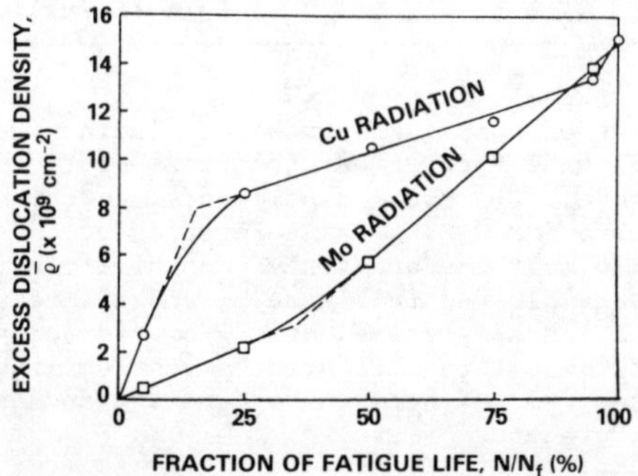

Figure 8 - Comparison of dislocation density measured with Cu and Mo radiation

To test the predictive capability of this method using the line-broadening method, specimens of aluminum 2024 were fatigued at four different stress amplitude blocks for varying number of cycles. X-ray line-broadening measurements were taken after each block using both copper and molybdenum radiation. From a calibration curve obtained from measurement taken at a single stress amplitude, as shown in Figure 8, the values of N/N_f were obtained. The amount of fatigue after cycling at each block is shown in Columns 5 and 6 of Table I. At the end of the fourth cycling block, the specimens were fatigued to failure at a stress amplitude of 280 MPa to determine the remaining fatigue life. For specimens A and B, the X-ray line-broadening method predicted that the remaining fatigue life was 45% and 54%, respectively, compared to the measured values of 42% and 58%. The agreement is considered to be excellent. When Miner's method (19) is applied, the remaining life should have been 20%. A method for calculating fatigue damage based on the role of the surface layer proposed by Kramer (20) was used to calculate the damage reported in Column 4. The agreement between the actual and calculated fatigue damage is good.

Table I. Comparison of Cumulative Damage Estimates for Aluminum 2024 Spectrum Specimens Under High-Cycle Fatigue Condition

Cycling Block	Number of Cycles and Stress Amplitude	Expended Fraction of Life			
		Miner	Kramer	X-Ray DCD	
				Sample A	Sample B
1	36×10^3 cycles at ± 170 MPa+	0.20	0.20	0.19	0.15
2	7.5×10^3 cycles at ± 210 MPa+	0.40	0.31	0.29	0.30
3	3.0×10^3 cycles at ± 245 MPa+	0.60	0.49	0.41	0.38
4	1.5×10^3 cycles at ± 280 MPa+	0.80	0.63	0.55	0.46
5	Cycled to failure at ± 280 MPa	Remaining Life			
	Sample A: $N_5 = 3.19 \times 10^3$	X-ray DCD		0.45	0.54
	Sample B: $N_5 = 4.36 \times 10^3$	Actual (N_5/N_f)		0.42	0.58

In another series of tests, specimens of aluminum 2024-T4 were fatigued in a solution of 3.5% NaCl for various fractions of the life at three different stress amplitudes. In these experiments, chromium radiation was used and a calibration curve for β_{mo}/β_{cr} as a function of the fatigue damage, N/N_f was obtained. To test the predictive capability using an extrapolation technique the data were plotted in Figure 9, and a linear relationship was employed to determine the N/N_f values at β_{mo}/β_{cr} for equal to one. For the specimen fatigued at these three different stress amplitudes (241, 276 and 310 MPa), the intercept ranged between 94% and 98% and indicates that the linear extrapolation procedure is valid.

Recently, W. Mayo (21) investigated the fatigue behavior of aluminum 2024-T4 under low-cycle fatigue conditions in the tension-tension mode ($R = 0.2$). Unlike the case of high-cycle fatigue, the dislocation density in the interior was essentially equal to that of the surface for measurement obtained after cycling in excess of 25% of the fatigue life. The dislocation depth profile was not measured at lower percentages. Similarly, a 1020 steel fatigued under low-cycle conditions in axial tension-compression ($R = -1$) displayed a similar behavior at $N/N_f > 25\%$ (22). Nevertheless, as the fatigue damage was increased, the β values increased in a linear manner, and failure occurred at a critical value of β^* that appeared to be independent of the strain amplitude (0.6% to 1.0%). A linear relationship of β/β^* was used to calculate the remaining life of specimens that were fatigued under block-loading conditions of various strain amplitudes. These data are shown in Table II, where again the agreement between the actual and the predicted cycles remaining is good.

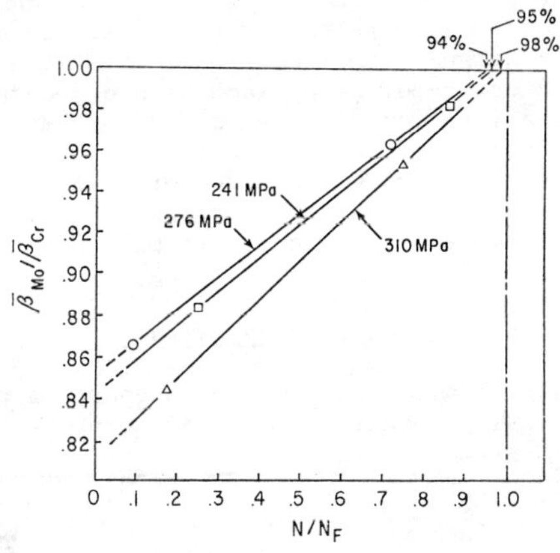

Figure 9 - Relationship between β_{Mo}/β_{Cr} and N/N_f for 2024-T4 Al in 3.5% NaCl

Table II. Comparison of Cumulative Damage for Aluminum 2024 Specimens Fatigued Under Low-Cycle Fatigue Conditions

Type of Loading	History	$\bar{\beta}$ Measured (minutes of arc)	Predicted Cycles Remaining	Actual Cycles
Monotonic	3,000 at 1.0% ε_{max}	22.09	4,550±1,550	6,430
Block	2,000 at 1.0% ε_{max} +10,000 at 0.6% ε_{max} +5,000 at 0.8% ε_{max}	19.65	26,025±5,225	27,285
Block	10,000 at 0.6% ε_{max} +2,000 at 1.0% ε_{max} +5,000 at 0.8% ε_{max}	22.09	22,230±3,230	19,842
Block	10,000 at 0.6% ε_{max} +5,000 at 0.8% ε_{max} +2,000 at 1.0% ε_{max}	18.89	26,600±3,230	22,730

The data for the prediction of stress-corrosion damage by X-ray technique is limited to 304 stainless steel exposed to 42% $MgCl_2$. An excellent correlation between the fraction of life time T/T_f and the ratio of the half width b/B was reported by Kamachi and coworkers (16), where b is the half width as a function of time, and B is the half-width of the original material measured at the surface (Figure 10). Note again, failure occurred at a constant value of b, independent of the applied stress. From Figure 6 it is also seen that the ratio of β_s/β_i would provide a measurement of the damage since failure would occur when $\beta_s = \beta_i$.

Acknowledgments

The authors gratefully acknowledge the support of this research by the David W. Taylor Naval Ship Research and Development Center.

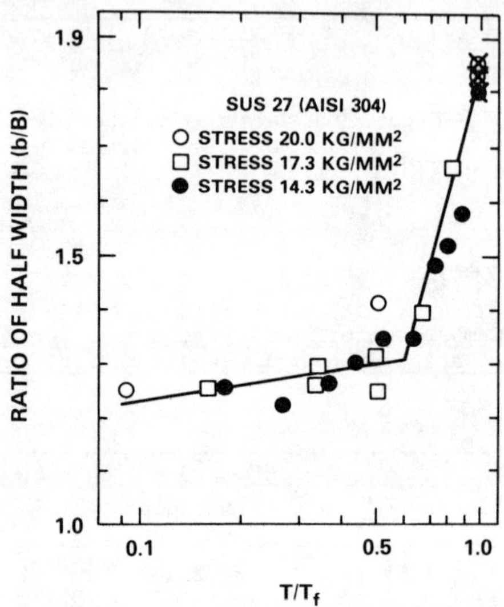

Figure 10 - Increase in the half-width as a function of the fraction of failure time T/T_f for 304 stainless steel subjected to stress corrosion in $MgCl_2$ at various stresses

References

1. I. R. Kramer and L. S. Demer, TMS-AIME, 221 (1961) p. 780.

2. I. R. Kramer, TMS-AIME, 230 (1964) p. 991.

3. H. Shen, S. F. Podleseck, and I. R. Kramer, Acta Met, 14 (1966) p. 341.

4. I. R. Kramer, Proc. Air Force Conf. on Fatigue, AFFDL-TR-O-144 (1969).

5. I. R. Kramer, Corrosion, 3 (11) (1975) p. 383.

6. I. R. Kramer, TMS-AIME, 5 (1974) p. 1735.

7. P. B. Hirsch, "Mosaic Structure," Progress in Metal Physics, 6 (1956) p. 236.

8. B. E. Warren and B. L. Averbach, Journal Applied Physics, 21 (1950) p. 595.

9. A. J. C. Wilson, X-Ray Optics, Methuen, London (1940) p. 37.

10. S. Taira and K. Hayaski, Proc. 9th Jap. Conf. on Testing Materials, 1 (1966).

11. S. Taira, K. Tanaka, and T. Tanable, Proc. 13th Jap. Congr. on Materials Research, 14 (1970).

12. S. Weissmann, R. Pangborn, and I. R. Kramer, Fatigue Mechanics, ASTM-STP-675, 163 (1979).

13. Yu. V. Baranov, E. P. Kostyukova and I. M. Makhmertov, Problemy Prochnosti (4) (Apr 1978) p. 483.

14. I. R. Kramer, TMS-AIME, 233 (8) (1965) p. 1462.

15. R. M. Yanici, S. Weissmann, and I. R. Kramer, Work in Progress (1979).

16. K. Kamachi, T. Otsu, and S. Obayaski, Journal Japan Inst. Metals, 35 (1971) p. 64.

17. I. R. Kramer and A. Kumar, Corrosion Fatigue, NACE-2 (1972) p. 146.

18. I. R. Kramer, Corrosion, 31 (1975) p. 383.

19. M. A. Miner, Journal Applied Mechanics, Series E-12 (1945) p. A-159.

20. I. R. Kramer, Proc. 2nd Int. Conf. on Mech. Behavior of Materials, 812 (1976).

21. W. Mayo, PhD Thesis, Rutgers University (1982).

22. I. R. Kramer, Work in Progress (1982).

APPLICATIONS OF SMALL-ANGLE NEUTRON SCATTERING METHOD

(SANS) IN NON-DESTRUCTIVE EVALUATION (NDE)

M. Fatemi, C. S. Pande, and B. B. Rath

Naval Research Laboratory, Washington, D. C. 20375 U.S.A.

Introduction

In the past decade, small-angle neutron scattering (SANS) has received considerable attention because of its role in providing nondestructively information on the shape and size distribution of microstructural defects in a variety of materials. Compared to other commonly employed NDE techniques, SANS has the particular advantage of detecting much smaller size defects which can be easily related to their sites of nucleation and consequently to the mechanisms for property modifications of materials.

These advantages are due to: (a) the use of neutrons with longer wavelengths (wavelength ranging between 5Å to 10Å) than X-rays (which eliminates multiple Bragg scattering phenomenon, and consequently simplifies the analysis of the observed spectra), (b) increased depth of penetration (commonly several mm) and, (c) the widely varying neutron scattering cross-sections with atomic numbers (permitting clearer distinction of scattering contributions from neighboring elements in the periodic table). Because of these advantages SANS can be used as an ideal tool for nondestructive study of voids, precipitates, and many other types of defects in solids.

Fig. 1 illustrates the practical detection limits of commonly used NDE techniques. It should be noted that all these techniques are either limited to near-surface defects with a resolution no better than 3-5 μm, or can only be used in bulk measurements for detecting very large defects (0.5 to 1.0 mm). In contrast, SANS method can be used to examine bulk defects in sizes ranging between 100 to 1000Å (0.01 to 0.1 μm).

It should, however, be recognized that the capability of small-angle scattering in flaw detection depends on the existence of a distribution of small defects with a relative high population, whereas other NDE methods can effectively detect isolated defects providing that their size range is within detection limits. A brief description of the principles and procedures for evaluating defect size and size distribution as well as a review of recent studies is presented in the following sections.

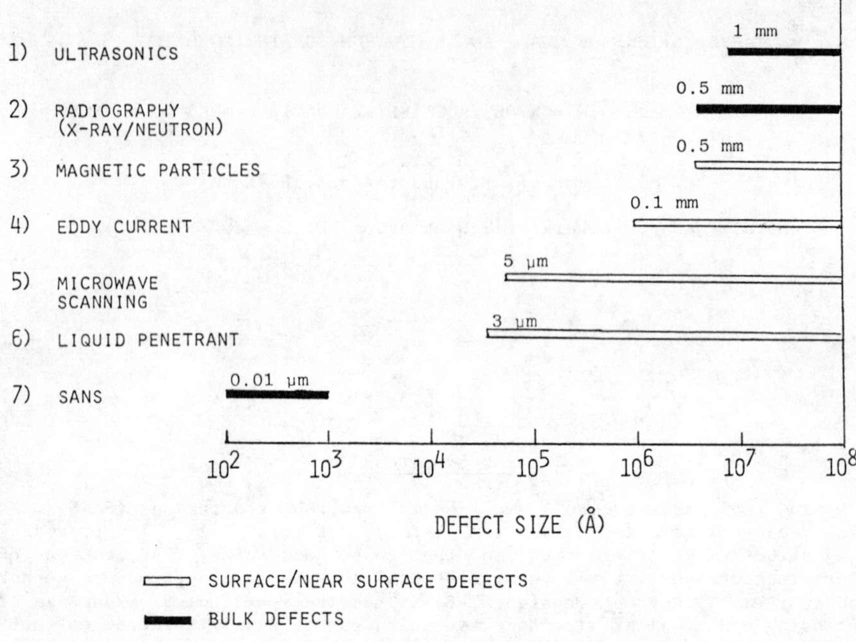

Figure 1. A comparison of detection limits of various NDE techniques. Shaded bars: bulk methods; open bars: surface methods.

Experimental Procedures

The source of neutrons for SANS studies is a research reactor from which the highest flux of cold neutrons with a broad spectrum of long wavelengths ranging between 4Å and 12Å can be obtained. As shown in Fig. 2, the essential components of SANS spectrometer may be classified into three general segments: (1) a system to provide monochromatic collimated neutrons, (2) the sample chamber and (3) the flight path-neutron detection system. The first segment is composed of: (a) the "cold source," which moderates the high-energy neutron at the reactor to slower velocities and longer wavelengths; (b) the curved neutron guide tube, whose function is to eliminate the unmoderated fast neutrons as well as background gamma rays; (c) the monochromator or wavelength selector and collimation system. Today, various SANS facilities employ one of two methods for wavelength selection, each with its own advantages and drawbacks. One method uses the graphite monochromator system, which is composed of several layers of graphite crystal wafers stacked at a slight angle (about 3 minutes of arc) to each other, in order to maximize the characteristic diffracted neutron output at 4.75Å. The second method uses a velocity selector, which is made up of a number of helical slots carved into the surface of a rotating cylinder. The angular velocity of the cylinder is chosen so that only those neutrons corresponding to a given mean wavelength (or average velocity) can pass unobstructed through the grooves. The wavelength resolution of a velocity selector depends on the incident beam spectrum, the selector geometry and the particular wavelength desired. Typical wavelength spreads range from 5% to 25% of peak wavelength depending upon the desired resolution and intensity.

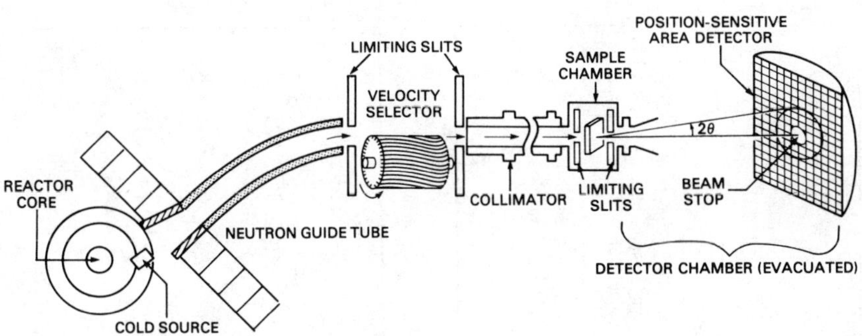

Figure 2. Schematic arrangement of components in a typical small-angle scattering spectrometer.

The main advantage of a graphite monochromator over a velocity selector are: (a) relatively high wavelength resolution of 5 to 10% and (b) appreciable reduction of background radiation to minimize the detector shielding problem. However, the method has the disadvantage in its lack of flexibility in wavelength selection in the requirement of bigger space. In contrast, the velocity selector allows a much wider selection of wavelengths, at lower wavelength resolution. The choice of one method of monochromatization over another is thus essentially governed by the availability of space in a particular reactor. A typical incident-beam collimator limits the beam size at the specimen to a maximum of 5-10 cm^2. Limiting slits may be used at both ends of the collimator to reduce the beam divergence as needed. The second unit consists of the sample chamber which is often evacuated together with the collimator and detector housing and may be equipped with automatic sample exchanges, hot stage and cryogenic devies, and eletromagnets. The third group of components in a SANS spectrometer is the flight path and the detector chamber which houses the position sensitive area detector. Detector chambers, including the evacuated flight path is commonly designed with provision for sample-to-detector distance adjustments. This permits selection of a wide range of scattering angles. A two-dimensional position sensitive detector grid usually consists of a 64x64 array of individual detecting channels, the radial position of which corresponds to a given angular distance from the main beam.

The detector output from a given measurement, is stored in a data acqusition system for processing and analysis. The simplest and perhaps the most descriptive approach of reviewing the experimental data is through an isointensity contour plot (see Fig. 3a), which is a plot of lines of equal intensity separated in either linear or logarithmic steps. In addition, data from rows or columns of detector channels of selected width may be processed and displayed as a plot of intensity vs channel number. Computer algorithms may be used to represent to data in triaxial isometric plots on the CRT screen. This type of display is quite useful in examining scattering anisotropies.

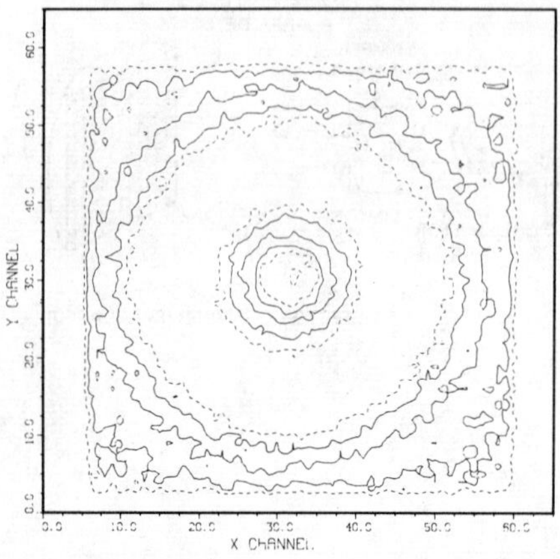

Figure 3a. Isotropic neutron scattering profile on a two dimensional (area) detector. Specimen: metastable beta titanium, slowly heated to 750°F/held 20 hrs.

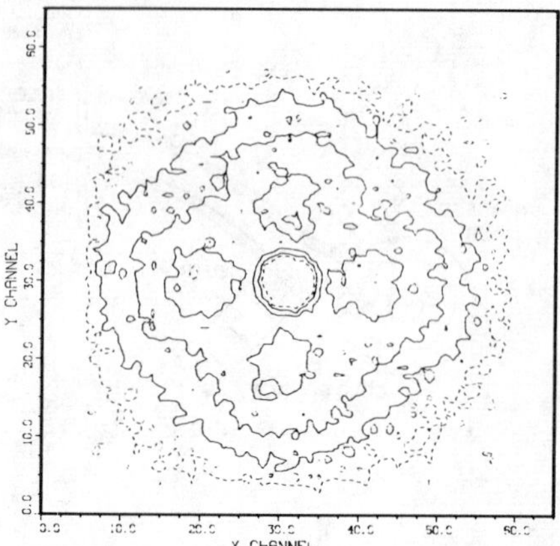

Figure 3b. Anisotropic neutron scattering from metastable beta titanium, rapidly heated to 750°F/10 min.

Isotropic contour plots can be immediately reduced to intensity vs angle plots by means of radial averaging, in which channels of a given radius on the detector - corresponding to a specific angle - can be summed and normalized. Anisotropic contour plots require other averaging procedures, such as rectangular and sector averaging. In the "rectangular" method, a narrow range of channels chosen in any specified directions is averaged for different angles. In the sector method, a wedge-shaped region of detector data is averaged from the center outward. This procedure may be repeated at various azimuth angles for a more complete analysis (Fig. 4).

Instrumentation - SANS Facilities Available

Because of their relatively high cost, there are only a limited number of SANS facilities available. Several SANS instruments have been described in the literature. Table 1 lists some of the most well known. The largest neutron scattering research center, the Institute Laue Langevin in Grenoble, France, is a cooperative center devoted mostly to the research supported by the European countries, viz, France, Germany and the United Kingdom. Other centers such as the ORNL (Oak Ridge National Laboratory) 30-meter SANS facility, the National Bureau of Standards research reactor and the University of Missouri Research Reactor (MURR) in

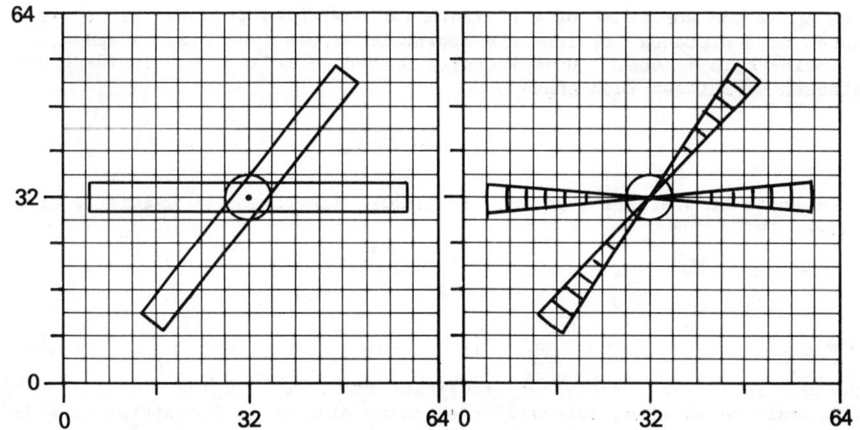

Figure 4. Two different modes of contour plot averaging for anisotropic data. (a) rectangular averaging; (b) sector averaging.

the United States, the Safari-1 SANS facility in South Africa, the Fiat Research Center in Italy and the Harwell small angle diffractometer in the U.K. have been in use for some time. Details of operation of these facilities may be obtained from individual centers as well as from appropriate references listed in Table 1.

Table I

SANS Facilities in Operation

		Reference
1.	Institute Laue-Langevin, Grenoble, France	1
2.	Oak Ridge National Laboratory, U.S.	2,3
3.	The National Bureau of Standards, U.S.	4
4.	The Fiat Research Center, Italy	5
5.	The University of Missouri SANS Facility, U.S.	6
6.	The Safari-1 Research Reactor	7
7.	The Harwell Small-Angle Diffractometer	8
8.	The FRJ-2 Reactor in Kernforschungsanlage, Julich	9

Principles of Small Angle Scattering

The technique of small angle scattering (SANS) as well as the related techniques using x-rays (SAXS) and light (SALS) are all based on the well known scattering phenomena described as:

$$A_k = A_o f_k \exp(-i\vec{q}\cdot\vec{\lambda}) \qquad (1)$$

where A_k is the amplitude of the radiation scattered from the point at r located at a distance |r| from the arbitrary origin (see Fig. 5) and A_o is the incident amplitude. Unit vectors S_o and S define the incident and scattered directions such that:

$$q = \frac{2\Pi}{\lambda}(S-S_o) \tag{2}$$

If the angle of scattering is 2θ, by simple geometry, the magnitude of q is given by:

$$|q| = \frac{4\Pi \ \mathrm{Sin}\theta}{\lambda}$$

since

$$|S| = |S_o| \tag{3}$$

Finally, f_k is the fraction of radiation scattered by the element at k which could be an atom, molecule or a volume element. The differences in the various small angle scattering (SAS) techniques are primarily related to the nature of f_k. For neutron scattering f_k is associated with the scattering nucleus, for x-rays with electron densities and for light with refractive indices.

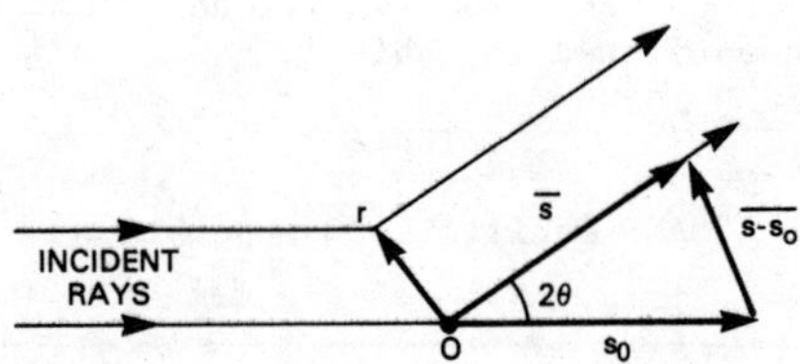

Figure 5. Parameter S, S_o and 2Θ defined.

SAS is generally considered as the phenomenon of scattering of radiation, at scattering angles, Θ, smaller than Bragg angles or at small q values. It is clear, however, that depending on the radiation used, the magnitude of θ can differ considerably. SAS range is, therefore, usually defined as

$$0 < |q| < \frac{\Pi}{d} \tag{4}$$

where d is the interatomic distance in the sample. Because of the short wavelengths, this range will be small for normal x-ray radiations, and large for neutrons.

The principles of SAS is essentially the same for all radiations and has been well developed for x-rays. The basic frame work leading to a mathematical description of scattered intensity is briefly presented here.

Intensity Due to a Particle

The total amplitude of the radiation scattered by a particle is the sum of scattering amplitudes, given by equation (1) over all k. The intensity I(q) is then obtained by multiplying ths sum with its complex conjugate. If it is assumed that: (a) all particle orientations are equally probable, (b) summation is replaced by integration over the average volume element, V, and (c) integral limits are extended from 0 to ∞ instead of confining only to the volume of the particle than one obtains.

$$\frac{I(q)}{I(o)} = \int_0^\infty 4\Pi V(r)p^2 \frac{\sin qr}{qr} r^2 dr \tag{5}$$

Where I(o) is the incident intensity, $\overline{V}$ is the average value of V, V is the volume of the particle, and p is scattering power.

Debye and Bueche defined a characteristic function $\delta_o = \frac{\overline{V}(r)}{V}$ which can be used to modify (5) as:

$$\frac{I(q)}{I(o)} = Vp^2 \int_0^\infty \delta_o \frac{\sin qr}{qr} 4\Pi r^2 dr \tag{6}$$

Equation (6) can be generalized to the form

$$I(q) = K\, \eta^2 \int_0^\infty \delta(_r) \frac{\sin qr}{qr} r^2 dr \tag{7}$$

which is the well known Debye and Beuche equation.(10)

Here K is a constant whose value depends upon the type of radiation, η^2, represents the mean square fluctuation in scattering power given by

$$\eta^2 = \Sigma\ (\rho_i - \rho)^2 \tag{8}$$

and where ρ is the mean scattering power of the medium or the matrix. As we have seen δ is a characteristic function (C.F.) in scattering power defined as

$$\delta(r) = \frac{\langle(\eta_i \eta_j)\rangle}{\eta^2} \tag{9}$$

where the symbol ⟨⟩ designates averages over pairs of volume elements separated by a scaler distance r. For a two phase system η^2 may be given as

$$\eta^2 = \ell_1 \ell_2 (\rho_1 - \rho_2)^2 \tag{10}$$

where ℓ_1 and ℓ_2 are the volume fraction of the phases having scattering powers ρ_1 and ρ_2. Debye and Bueche's equation (7) is the most general approach to small angle scattering of any type of radiation. The small angle scattering intensity is unambiguously given by the characteristic function $\delta(r)$. Thus, any information contained in the SAS intensity profile is implicitly contained in C.F.

With the aid of this function, the theoretical methods for the evaluation of small angle scattering of dilute systems (i.e., widely scattered particles, etc.) of identical particles have been well developed. For a great variety of particles of simple shapes such as spheres or ellipsoids, oriented at random, exact C.F. are known, and predicted intensities using (7) have been shown to be in agreement with experiments. For systems where particles exhibit preferential orientations, equation (7) may be generalized such that $\rho(r)$ is an angularly dependent function and the integration is then done not only over the magnitude but also the angles of the vector r. The SAS intensity will then depend on q as well as on the azimuthal scattering angles.

<u>Radius of Gyration</u>

Although quantitative evaluation of C.F. is sufficient to completely understand the SAS profiles and this C.F. can be obtained from a Fourier inversion of the scattering data, it has intuitively no simple connection either with the size or the shape of the particle. In practice one can choose some 'integral' parameter (integration of some parameters over C.F.) and regard it as a measure of particle size. Radius of Gyration introduced by Guinier is one such 'integral' parameter.(11)

If the sin qr/qr term in equation (7) is expanded in a series one obtains

$$I(q) = I(o)\ [1 - \frac{R_g^{\ 3}}{3} q^2 + \ldots] \qquad (11)$$

where I(o) is the intensity at q = 0, R_g^2 is the mean square radius of gyration of the scattering object defined as:

$$2R_g^2 = \frac{{}^{\infty}\!\int_o \delta(r) r^4 dr}{{}^{\infty}\!\int_o \delta(r) r2dr} \qquad (12)$$

By analogy with classical mechanics, R_g is referred to as the radius of gyration. If higher order terms in q are included in (11) a better approximation for I(q) is

$$I(q) = I_o\ (\frac{-q^2 R_g^{\ 2}}{3}) \qquad (13)$$

It should be noted that equation (13) can be obtained without introducing the characteristic function. Furthermore, this equation is exact up to 4th power of q and is also valid for q approaching ∞. (It is valid for all q provided qRg < 1.0.) R_g is conveniently obtained from a slope of a plot of ln I(q) vs. q^2 which would be a straight line based on equation (13). Such a plot is often referred to as the Guinier plot and has been extensively used in characterizing defect sizes in many systems. However, systems containing identical particles with well defined shapes, a basic premise of the above formulation, are rarely found in practice. In such cases, the Guinier plot will show a departure from straight line with the deviation increasing with the increasing q. The more anisometric a particle, the more will be the deviation from linearity. A very similar effect is caused if the particles are of different sizes (polydispersity). In some cases the effect due to anisotropy that due to polydispersity could be identical. For example, Guinier (11) has noted that the scattering intensity of identical ellipsoids are identical to that of an assembly of sphere with a special size distribution. In short the determination of the shape and polydispersity from SAS is never unambiguous. However, making reasonable assumptions of size distribution, meaningful results can be obtained.

Porod's Law

The characteristic function, as defined earlier can be expanded as a power series such that

$$\delta(r) = 1+a_1r+a_2r^2+a_3r^3+\ldots. \tag{14}$$

It can easily be shown (11) that a_1 = S/4V where V is the total volume of all the particles and S their total surface area, i.e. Using the series expansion above, one can integrate (12) and obtain I(q). It will be seen that only the odd powers in r contribute substantially.

$$I(q) = \frac{4\Pi}{q^4}\left(\frac{-2\ S}{4V} + 4\frac{a_3}{q^2} - 6\frac{a_3}{q^2}\right) \tag{15}$$

(plus a small oscillating terms which will be canceled out unless the system is too regular). Therefore for large angles ($qR_g \gg 1$)

$$I(q) \rightarrow \frac{2\Pi}{q^4}\ \frac{S}{V} \tag{16}$$

This is the so-called Porod's law (12). The equation was obtained independently by Porod and Debye.

Characteristic Function

As pointed out earlier, the characteristic function (C.F.) gives the intensity scattered unambiguously. C.F. has been calculated for many systems. For example, Debye, Anderson and Brumberger (13) have shown that for a randomly dispersed system the C.F. is given as

$$\delta(r) = \exp(-r/a) \tag{17}$$

where a is a correlation distance and is a measure of the size of the fluctuation. Substituting this value of C.F. the intensity I(q) is obtained as

$$I(q) = (K/\eta^2)a^3(1+q^2a^2)^2 \tag{18}$$

This gives the scattered intensity for a whole range of q values. For large q, it predicts

$$I(q) \rightarrow \frac{1}{q^4}$$

In accordance with Porod's law, C.F. thus can be of great use in interpreting scattering from metallurgical specimens.

Scattering From a Group of Particles

So far we have been mainly concerned with scattering by a single particle. For practical applications one is more likely to be interested in low angle scattering from a group of particles. When the system is very dilute (widely separated particles), and all the particles can with equal probability take all possible orientations, the scattered intensity from a number of particles N can simply be assumed to be just N times the intensity from a single particle. Numerical methods are now available for obtaining particle size distribution from intensity profiles. Here, one, of course, assumes that the orientations are completely random and are in no way related to the position of any two particles. When the above assumptions are no longer valid, and especially when the particles are close together, each individual case has to be treated separated. The SAS intensity for closely packed particles may exhibit a peak instead of monotonically decreasing with q. The area under the peak, and q_{max} (q value at the peak) then contain important information about the size and volume fraction of the particles. For example, Pande (14) has shown that q_{max} may approximately be related to the volume fraction F_v of the system by the following relation

$$\frac{\Pi}{q_{max}} = R\ [1 + \{\exp\ (8Xv)/6Xv^{1/3}\}\{\Gamma 1/3,\ 8Xv\}] \tag{19}$$

where 2R is the size of the defect assumed approximately spherical and $\Gamma(1/3,\ 8Xv)$ is incomplete Γ function of argument (1/3, 8Xv).

Babinet Principle of Reciprocity

In many metallurgical systems one is interested in characterizing the size and volume fraction of voids, rather than precipitates. One, however, does not need to calculate ab initio the scattering for such a system, but simply take over the results for that already obtained for particles, or precipitates. The fact that this is possible is due to the validity of a certain principle usually known as Babinet principle of reciprocity which states simply says that complementary objects produce the same diffraction effects. A system containing a distribution of precipitates, is complimentary to a system containing voids, where the precipitates were present. In mathematical terms if $\rho_1(r)$ is the electronic density it will be complementary to a space with electronic density $\rho_2(r)$ if $\rho_2 = \rho_o - \rho_1(r)$ where ρ_o is a constant. This can be seen from equations (7) and (10).

Extension to SANS

The results obtained so far although historically obtained mostly for SAXS can be extended to SANS in toto. For this purpose it is necessary to remember as stated before, that in SANS the scattering of the incident radiation (neutrons) are the atomic nuclei. The coherent scattering cross section $d\sigma/d\Omega$ for neutrons is given by the well known equation (15)

$$\frac{d\sigma}{d\Omega} = \frac{1}{N_o}\left(\frac{M_n}{(2\Pi h)^2}\right)^2 \left|\int V(r)\ \exp(i\ q.r)d^3r\right|^2 \tag{20}$$

where σ = total scattering cross section
N_o = incident neutron flux
V = scattering potential
$d\Omega$ = solid angle over which the scattered neutrons are collected
M_n = mass of the neutrons.

The above equation leads to the concept of scattering length b. Consider that the scattering system is just a single atom. In this case, neutron wavelengths are larger than nucleus dimensions $|q.r|<<1$ and one can write

$$\frac{d\sigma}{d\Omega} = \frac{1}{N} \left(\frac{M_n}{(2\Pi h)}\right)^2 |\int V(r)e^3r|^2 = \frac{1}{N} b^2 \tag{21}$$

where b is called the scattering length and defined as (16)

$$b = -\frac{M_n}{2\Pi n} \int V(r)d^3r \tag{22}$$

For a system containing many atoms in a volume V, $d\sigma/d\Omega$ is per unit volume is given by

$$\frac{d\sigma}{d\Omega} = \frac{1}{V} {}_r |b_k e^{i\vec{q}.\vec{r}}|^2 \tag{23}$$

where b_r is the coherent scattering length of the particular atom occupying a site with a position vetor r, and V is the volume of the sample exposed to neutrons. For many atoms close together, since SANS can only detect inhomogenities much larger than atom size, it is permissable to replace the scattering length with the locally averaged scattering length density $\rho(r)$ where $\rho(r) = b(r)/w(r)$ where w(r) is the volume of the atom at r. Then on changing the summation to integration over the volume V of the specimen we get

$$\frac{d\sigma}{d\Omega} = \frac{1}{V} |\rho(r) e^{i q r.} d^3r|^2 \tag{24}$$

If I(q) is identified with $d\sigma/d\Omega$ and $\rho(r)$ is identified with electronic density ρ the mathematical description for SAXS discussed before can be taken over for all the SANS work (except magnetic scattering which are due to special properties of neutrons).

Typically $b \sim 5X10^{-15}$m, and
$\rho \sim$ a fraction of $10^{15}/m^2$

Unlike x-rays, b or ℓ does not vary monotonically with atomic number Z and atoms with Z very close to each other may scatter very differently.

The parameter $d\sigma/d\Omega$ is related to the measured parameters by the following relation

$$[I_m(q)] = I_mo \ t \ _AT \frac{d\sigma}{d\Omega}(q) \ \Delta \ \Omega + I_\infty \tag{25}$$

here $I_m(q)$ = number of neutrons scattered into the detector corresponding to the wave vector q,

$\Delta\Omega$ = solid angle subtended by the collector at the specimen,

t = time of collection,

T = specimen thickness typically in mms,

A = area of specimen exposed to beam; typically 10mm X 10 mm,

I_∞ = number of neutrons incohsently scattered into the detector, i.e., $I_m(q)[q\to\infty] = I_\infty$.

Numerical Methods for Obtaining Defect Size Distribution

Once the intensity scattered from a system containing a distribution of defects have been measured for all possible values of q, one is often interested in inverting this data to obtain the size distribution of the defects or scatterers. Early methods used a number of simplifying assumptions and were only partially successful. After the advent of computers a number of successful numerical techniques have been developed to carry out this inversion rigorously at least for dilute systems. It is still essential to have some information regarding the nature of the scatterers. This is usually obtained by TEM or some other similar method. Some extrapolation to other q values not measured are also sometimes required. We briefly describe some of these numerical methods. More details can be obtained from the references cited below.

As seen before, the general form of small-angle scattering intensity as a function of the wave vector q may be expressed as

$$I(q) = I_o \int_{R_{min}}^{R_{max}} N(R)\ \phi\ (qr)\ m^2(R)dR \qquad (26)$$

where I_o is a constant related to the incident beam flux and detection geometry, N(R)dR is the number of particles in the range R and R+dr, $\phi(qR)$ is the intensity shape function and $m^2(R)$ is the square of the excess nuclear scattering length.

The various size distribution algorithms in use today employ one of the three procedures for the inversion of the integral in equation (26).

Glatters Method. In this method, N(R) is described in a series of cubic spline functions. Upper and lower limits of integration in (26) are assumed, meaning that $R_{min}<R<R_{max}$.(17)

Vonk's Method. The expression of equation (26) is converted to a sum by dividing the range of integration into a large number of intervals. The resultant matrix equation is solved in the standard fashion for several forms of $\phi(qR)$. Constraints must be placed on the solution in the form of damping functions to prevent large oscillations near end-points.(18)

Mellin Transform Method (Mook)(19), (Walther and Pizzi)(20). Here the integral in (26) is solved using the Mellin transform assuming spherical particles with Rg as the effective radius of gyration. This method requires the extension of intensity data to $q\to\infty$. This means that the wide angle behavior of the data should be known fairly accurately. However, because of low neutron counts and poor statistics at high angles, calculated size distributions may not fully agree with the expected results.

Applications of SANS to Metallurgical Problems

Experimental and theoretical research on SANS applications has made significant advances in the past few years. Most important applications of SANS to metallurgical systems consist of studies relating to cavitation, precipitation, phase transformations and to a lesser extent to dislocations. Majority of the technologically relevant applications have been in studies of voids and precipitates.

Microstructural defects in material form either during processing fabrication such as casting, forging and rolling, superplastic forming, powder compacting, heat-treatments and quenching, or during service life such as fatigue, creep, radiation damage and high temperature aging. In addition, the resulting size and distribution of the defects depend upon the alloy system under consideration. SANS can be used to characterize most of these defects. Furthermore, it can also be used to study magnetic materials through their specific magnetic scattering which must be clearly distinguished from defect related scattering.

In this paper, in order to illustrate the SANS application, two metallurgical problems, i.e. nucleation and growth of voids and solid-state transformation are discussed in detail. Many other applications are briefly mentioned and a detailed but selected list of recent applications is provided in Table II. This table lists the investigation objective, material system and main findings and illustrates the versatility of the method for detection of a wide variety of material defects. Further details may be obtained from the references provided in this table. (See also reviews by Weertman (56) and Gerold and Kostorz (57).

Precipitation and Phase Transformation in Solids

SANS applications to the precipitation phenomena fall into two major categories: a) dilute systems, in which the concentration of precipitates is no more than 0.5% and b) densely-packed systems containing large volume fractions up to 80%. Although SANS research was initially concerned with problems of the first type, in recent years increasing emphasis is being given to the study of second type of problems.

As stated previously, size distribution of defects in dilute systems is determined by the use of one of several algorithms, assuming a simple particle shape. However, when the particle density increase beyond a critical volume fraction leading to interparticle effects or gross deviation from simple geometrical shapes, the two-dimensional scattering profile exhibit anisotropies. In such situations the use of simple particle size distribution algorithm is no longer void.

To assess the utility of small-angle neutron scattering (SANS) as a non-destructive method applied to the phenomenon of high-density precipitation, systematic SANS and TEM studies have been done on a model system, viz ω precipitates in a Ti alloy. This study demonstrates the applications of SANS to a metallurgical system, undergoing phase transformation and to bring out some new techniques developed for such analysis. This work is now described in some detail.(58)

The choice of beta-III titanium alloy (Ti-11.5Mo-6.0Zr-Sn-0.14O_2) was based on the apparent ease of controlling the size distribution of omega phase precipitates, and was also due to the feasibility of approximating the ellipsoidal morphology to spherical for certain simplifications desirable in the SANS analysis. The isothermal omega phase has an hcp

Table II

SELECTED METALLURGICAL PROBLEMS INVESTIGATED BY SANS

A. Precipitation and Phase Transformation Phenomena

Material System	Investigations Objective	Processing Variables	Defect Type/Size	SANS* Facility	Remarks	Ref. No.
Al_2O_3-S_iO_2 System	Phase Decomposition	Extrusion, Air Anneal.	Decomposition, Large Density fluctuation of Si, Al	J	Abstract only	21
Al-Si Alloys	Identification of SAS sources	Irradiated $5x10^{18}/cm^2/sec$	Precipitates	L	Single x'tal $q=.007 \rightarrow 0.03 Å^{-1}$	22
Al-6.8% Zr	Decomposition Kinetics	Homogenized 340°C Slow cooled 240°C aged 135°,110°,80°C	"Solid Solution G.P. Zones"	L	_In Site_ SANS	23
Al-Zn-Mg	Metastable Miscibility Gap	RT anneal/1Mo	Metastable G.P. Zones/30Å	L	Detector Distance; 70-240 Cm $\lambda = 6.6Å$	24
Cu-1% Co	Single to Multi domain transition of precipitates	600°C Aging for times up to 500h	Co-precipitation in Cu/200$\pm$20Å	J	2.5K Oe magnet $q < 0.04Å^{-1}$	25

L = Institute Laue Langevin
J = Julich
O = ORNL, Oak Ridge
H = Harwell

Table II

SELECTED METALLURGICAL PROBLEMS INVESTIGATED BY SANS

A. Precipitation and Phase Transformation Phenomena (continued)

Material System	Investigations Objective	Processing Variables	Defect Type/Size	SANS* Facility	Remarks	Ref. No.
Cu-1% Co	Single domain precipitates	"	Co-precipitates	J	$\lambda=9\pm1.8$Å uniform magnetization within particles; elongated ellipsoids	26
Na_2O-Al_2O_3-SiO_2-TiO_2	Atomic density fluctuation, growth kinetics	Annealing at 750°C 2-80 hrs	Size ~ 30-40Å		q=0.02 to 0.10Å	27
Invar alloys single crystal $Fe_{70}Ni_{30}$	Magnetization inhomogeneity, Electron irradiation effects	electron irradiation $4x10^{19}$ electron/cm^2 at 200°C	Chemical Clusters	L	Irradiated specimens show a pseudo-periodical magnetization fluctuation	28
Mn-Cu Alloys	Decomposition on aging	Annealing Temp 350, 450, 550°C 2 hours	Porod radii: 30-60Å	O	At 450°C the principal mode of decomposition is precipitation of small metastable zones	29

Table II

SELECTED METALLURGICAL PROBLEMS INVESTIGATED BY SANS

A. Precipitation and Phase Transformation Phenomena (continued)

Material System	Investigations Objective	Processing Variables	Defect Type/Size	SANS* Facility	Remarks	Ref. No.
Nickel Superalloy 713LC	Microstructural changs in creep	Temp: 923K 1123K Creep test σ=135-215 N/mm^2	γ-precipitates $M_{23}C_6$ carbide	J	Variation of anisotropy factor can be used as a measure of accummulated damages	30
Silicon hydride (SiH)	Morphology as a function of heat treatment	Heat treatment 375°C 10-15 min 575°C 25 min	Columnar structure Column diameter 60°Å	L	Matrix scharating the columns essentially voids.	31
S_iO_2-Na_2O Glasses	Phase separation kinetics	560°C, time	density and concentration fluctuation	L	q up to 0.05Å scattering peaks at ~ $1.4x10^{-2}$	32
S_iO_2-M_2O Glasses (M = Alkali metal)	Concentration fluctuation		Coherence length of fluctuation ~7Å	L		33

Table II

SELECTED METALLURGICAL PROBLEMS INVESTIGATED BY SANS

B. Voids

Material System	Investigations Objective	Processing Variables	Defect Type/Size	SANS* Facility	Remarks	Ref. No.
Al Single Crystals	Defect distribution	Neutron irradiation $2x10^{20}/cm^2$	Voids	0	λ=2.5Å Ge Mono-chromators void size dist. calculated	34
Al (High purity)	Void morphology, swelling	fast neutron flux	Voids 100-800Å	J	Detector at 2.0, 20, 12 m.TEM comparison	35
Cu-Single Crystal	Fatigue in Cu single-xtal		Voids			36
Alloy-800	Creep		Voids	L		37

Table II

SELECTED METALLURGICAL PROBLEMS INVESTIGATED BY SANS

B. Voids (continued)

Material System	Investigations Objective	Processing Variables	Defect Type/Size	SANS* Facility	Remarks	Ref. No.
High purity Copper	Cavitation at high temperature fatigue	fatigued, .07% to 25% of fatigue life.	Pores ~ 350Å	L		38
Pure Cu	Fatigue Induced Voids	No. of cycles, Temperature, stress	Voids	L	Voids 10-50Å predominantly along grain boundary	39
GaAs	Study of Scattering anisotropy	Fast neutron irradiation 25°C to 718°C	Oriented displacement spikes	L	Detector distance = 2.5m, 5.5m $q = 0.02$ to $0.10 Å^{-1}$	40
Nimonic Alloys 105	Temperature dependence of microstructure; degradation of creep resistance	a) 200-300 MPa/750°C b) 500 MPa/800°C c) as heat treated	600-800Å tentatively microvoids	L	Peak at 0.005Å sample to detector 5-20m.	41

Table II

SELECTED METALLURGICAL PROBLEMS INVESTIGATED BY SANS

C. Magnetic Alloys and Microstructure

Material System	Investigations Objective	Processing Variables	Defect Type/Size	SANS* Facility	Remarks	Ref. No.
Au-Fe Alloys	Ferromagnetism	Temperature; Fe Concentration	Critical Scattering	L	λ = 6.7Å $\Delta\lambda$ = ±8%	42
CoGa Alloys	Super paramagnetic assemblies in CoGa	Temperature 4.8K and higer fields up to $17KO_e$	Assembly size ~ GoÅ	L, H		43
E_rCo_2	Magnetic origin of scattering	Temp 10K to 150K field: 0 and 5 KOe			Scattering very dependent on temperature	44
Iron Nickel Invar alloy $Fe_{65}Ni_{35}$	Temperature variation of the magnetic cluster structure.	25°C to 422°C	Clusters		n=0.298 $Rg^{2.34}$ n=no. of atoms in a paramagnetic cluster, Rg= Guinier radius	45
Amorphous Fe-Ni-Mo-B	Magnetic & nuclear homogeneity	single-roller quenching cooling rate 10^6 C/sec.	R_g= 230Å (magnetic) R_g= 200Å (nuclear)	L	anisotropic without magnetic field λ=9.26Å S-D distance: = 2,87 m q=.007 to .065$Å^{-1}$	46

Table II

SELECTED METALLURGICAL PROBLEMS INVESTIGATED BY SANS

C. Magnetic Alloys and Microstructure (continued)

Material System	Investigations Objective	Processing Variables	Defect Type/Size	SANS* Facility	Remarks	Ref. No.
$HoMo_6S_8$	Long range magnetic structure	T_M=0.71-0.75K	Periodic structure in superconducting state	L	q=0.002-0.63Å^{-1} λ=6-15 Bragg-like peaks abar q=.03Å^{-1}	47
Fe65 (Ni1-xMnx)35	Ferromagnetic regions in antiferromagnetic matrix	Alloying variables Ni=24,25,30%	350Å and 100Å	not given		48
Fe-based metallic glasses	Chemical homogeneity and magnetic domain walls	Applied field: 4-10 KOe; annealing conditions: 350-380C, 45-60 min	Domain walls	O	Limited phase separation and chemical inhomogenity	49
Ni_3Mn	Magnetic properties in ordered system		Small, ordered clusters at T_c=750°; Magnetic clusters of T_c=400K, 40Å	J		50

Table II

SELECTED METALLURGICAL PROBLEMS INVESTIGATED BY SANS

C. Magnetic Alloys and Microstructure (continued)

Material System	Investigations Objective	Processing Variables	Defect Type/Size	SANS* Facility	Remarks	Ref. No.
$(Pd_{99.65}-Fe_{0.35})$	Neutron scattering as a function of temperature	Temperature 0–15K		Biology Facility at BNL	No spin waves observed within the ferromagnetic	51
$Tb_{.25}Cu_{.75}$	Magnetic properties of amorphous alloys magnetic short range order	Heat treatment 210C, 300C	Microscopic structure in local magnetization	CEN SACLAY France	Magnetic short-range ordering occurs at much higher than Currie temperature	52

Table II

SELECTED METALLURGICAL PROBLEMS INVESTIGATED BY SANS

D. Miscellaneous

Material System	Investigations Objective	Processing Variables	Defect Type/Size	SANS* Facility	Remarks	Ref. No.
Ag-Ge Liquid Alloys	Test of heteroatomic chain model	Ge concentration 940→1177°K		L	λ=4.9-7.11	53
Amorphous SiO_2	VSAS neutron irradiated amorphous Silica	Irradiated $3x10^{19}$ at 48°C	$R_G=3X10^5$Å inhomogeneities	J	Very small angles	54
Glass	Effect of surface imperfections	Polished Abraded		L	q=.005 to .025$Å^{-1}$	55

structure, and the precipitates are oriented in the bcc beta matrix such that $[111]_\beta \parallel [0001]_\omega$ and $[110]_\beta \parallel [2110]_\omega$. Four variants of the omega phase oriented in the possible [111]-direction have been observed. A second form of the ω-phase ("as-quenched") was also reported, produced during rapid quench from the beta phase. (for a recent review see ref (59).

The selection of proper aging times and temperatures for SANS study was based upon the $\beta \rightarrow \beta + \omega$ time-transformation temperature (TTT) diagram. The transformation temperature range was suggested to be between 300°C and 425°C. Five specimens were therefore aged at temperatures from 315°C to 425°C and at aging times ranging from 2 minutes to 100 hours. All measurements reported here were performed at the Oak Ridge National Laboratory High Flux Reactor 20-m SANS facility (3). The instrument permits variation of sample-to-detector distance up to about 19 meters, uses a graphite monochromatised neutron beam of 4.75A (0.475 nm) wavelength, and is equipped with a 64x64 element position sensitive area detector.

When the area detector data were radially-averaged, all profiles exhibited an interparticle interference peak as their major component. Since the peak position (q_{max}), which is a function of both interparticle distance and particle density, showed considerable coordinate shifts, three sample-to-detector distances of 2.5, 7.0 and 17.0 metres were chosen to cover all q ranges of interest. A selected number of specimens were extensively examined by transmission electron microscopy (TEM) for their size, range and morphologies.

SANS contour plots, representing the data on a two-dimensional array, consisted of two varieties. One group was nearly isotropic, the other distinctly anisotropic. Thus, the radially averaged plots of anisotropic data would show peaks at locations which nearly correspond to the overall interparticle effect.

An examination of SANS profile (see, for example, Fig. 6) obtained for the five temperatures reveals several important points. (1) In general, SANS profiles progress towards decreasing q's as the temperature increases. Thus, much wider profiles are obtained at 315°C than at, say 400°C. (2) At any temperature, the profiles may be divided into two groups, one corresponding to aging times of less than 1 or 2 hrs and characterized by peaks positioned at relatively low q_{max}, the other corresponding to longer aging times and larger q's. In the first group, only the interference peak intensity appears to change, but any shift in its position is difficult to ascertain near the limits of experimental resolution. In the second group the shift in q_{max} towards zero with increased aging time can be easily observed.

For a detailed analysis of the transformation phenomenon, SANS and TEM results from the same specimen were compared and analyzed to establish the correlation between the peak position (q_{max}) from SANS and size-related parameters from TEM, and also between SANS anisotropy and precipitate morphology. It is seen that SANS profiles with peaks at q_{max} are obtained from all the specimens irrespective of the morphology of the precipitates. The interparticle distance can be roughly correlated with the q_{max} by the relation $D_{int} = 2\pi/q_{max}$, provided the precipitates are sphere like. We found that D_{int} showed a consistent trend with temperature and time for ellipsoidal particles.

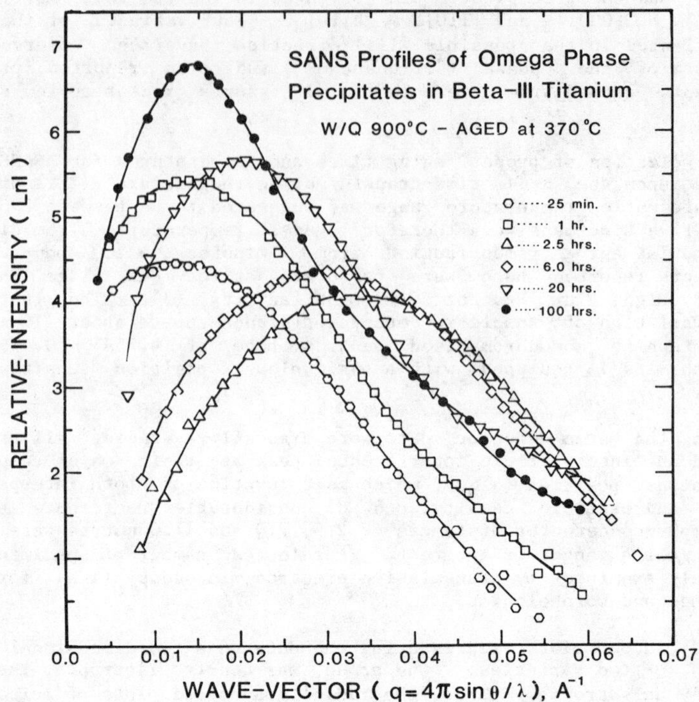

Figure 6. Intensity profiles for beta-III titanium isothermally processed at 700°F for times indicated.

The average size of ellipsoidal precipitates, as derived from TEM particle size histograms was also found to increase with both aging time and temperature. The volume fraction F_v of the precipitates remains relatively constant for specimens aged beyond about 2 hrs. Similar results were found by Hickman in Ti-25% V alloys aged beyond 5-10 hrs.(60)

It is instructive to examine the relationship between SANS experimentally obtained D_{int} and the observed TEM average size, i.e., whether in fact the two parameters are proportional. Fig. 7 shows a plot of the variation of SANS interparticle distance vs TEM particle size, and clearly indicates the proportionality between the two quantities in agreement with equation 19. Within the expected experimental variations, all points fall very nearly on a straight line.

The correlation between precipitate morphology and contour plot anisotropy is another important result of our SANS-TEM comparison. Specimens aged by 'rapid' heating for less than one hour always showed strong anisotropy, while those aged after 'slow' heating exhibited little or no anisotropy. The degree of anisotropy was found to be related directly to the aspect ratio as found from TEM. Thus, the 370°C/10-min. anneal specimen containing some of the longest omega needles ever encountered showed more anisotropy than the specimen with shorter needles.

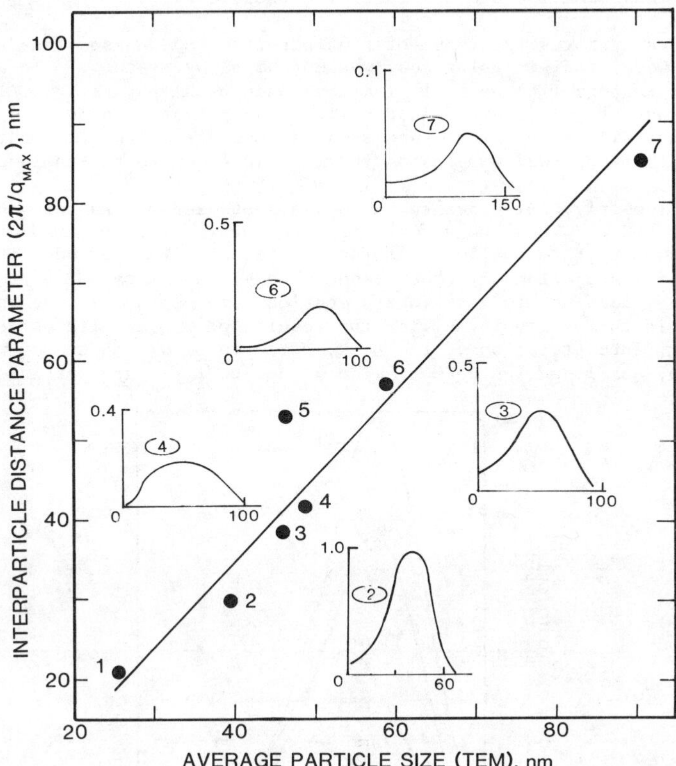

Figure 7. Plot of average interparticle distance from SANS peaks vs particle size observed from TEM (58).

The correspondence between SANS profiles, SANS contour plots and transmission electron microscopy in the study of size distribution and morphology of the omega phase precipitates has thus been established. In particular it was found that:

(1) The peaks in SANS profile are due to interparticle effects of precipitates, independent of the shape of the precipitates.

(2) The isotropic contour plots, and SANS profiles in which q_{max} showed a shift with changes in aging treatment, result from changes in the size of the 'sphere-like' ellipsoidal precipitates with aging treatment.

(3) Anisotropic contour plots and SANS profiles in which q_{max} did not substantially change with aging treatment result from the growth of rod-like precipitates.

(4) The degree of anisotropy was found to be directly related to the aspect ratio of the precipitates.

We now briefly mention some other applications of SANS in the study of phase transformation and precipitation phenomena.

The early investigations of precipitation and phase transformation phenomena were studied using mostly aluminum alloy systems. In a recent review by Kostorz (16) several Al-alloys such as Al-Zn, Al-Mg, Al-Si and Al-Mg-Zn have been presented in detail. The primary reason for choosing aluminum and its alloys for such studies was its high transparency to neutrons. This allowed samples as thick as 10-15 mm to be examined.

More recently, Al-6.8at%Zn alloy was studied by Raynal (61) and Messaloras (62). They showed clearly the progression of SAS peaks at q_{max}=0 from larger to smaller q's for different aging periods (See Fig. 8). Later investigations by other techniques TEM and x-rays SAS have also been made, leading to an interpretation of spinodal decomposition mechanism in this system. As with the results on ω precipitates described previously, interpretation of the peak, after in situ SANS measurements on hot stage, was found to be most likely due to the interparticle effect (62).

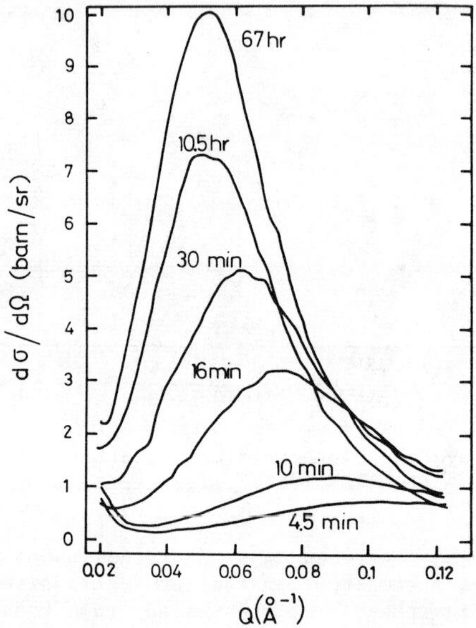

Figure 8. Scattering intensity for an Al-6.8% Zn alloy quenched from 310°C, aged at room temperature for times indicated (62).

Boeuf et al (63) studied the coarsening of precipitates in a 304 stainless steel. The observed SANS data followed a simple profile without peaks or anisotropies. The size distribution calculations were made under the assumption a bimodal distribution results of which agreed relatively well with TEM observations. The Fiat group (20) has for some time been active in the evaluation of the defects in engineering materials such as Nickel Superalloys (δ' precipitates), steels and non-ferrous materials. SANS results on δ' precipitates has shown the simultaneous coarsening of the precipitates and stability of the corresponding volume fraction for Inconel X-750. An example of the interparticle effect in structural materials may be seen in the SANS curves obtained for Udimet specimens at small q values, $q < .007Å^{-1}$ (Fig. 9). Other well-known application of

SANS by the Fiat investigators include measurements on gas turbine blades, beam broadening by carbon steels, carbide precipitation in austenitic steels, and to a smaller extent Al-and Cu-alloys and polymers.

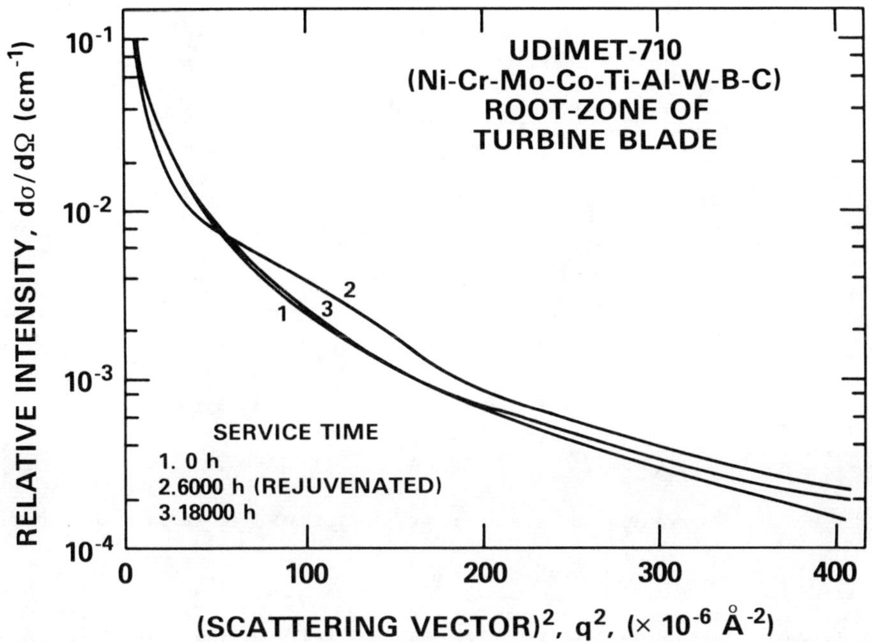

Figure 9. Coarsening behavior of γ' precipitates in Udimet 710 observed by SANS. Some interparticle effect is visible in curve 2 (20).

Osamura et al (64) have studied the effect of alpha-phase precipitation on the critical current density in a superconducting composite wire (Ti-27Nb-6Ta-6Zr(at%) using time-of-flight SANS. In these measurements, characteristic bumps (see Fig. 10) associated with interparticle interference were observed. From a calculation of the total integrated intensity and the Guinier radius, it was shown that the precipitates increased in size without changing their total number after only 10 seconds of aging at 643 K. These SANS measurements were consistent with previous findings that at intermediate magnetic fields the superconducting critical current density increases with the progression of α-phase precipitates, and that the relaxation of the dislocations plays only a minor role.

In addition to ω phase precipitation discussed before, SANS investigations on precipitation have been carried out at NRL on a large number of metallurgical systems (65). In an early work, the evolution of Nb carbon nitride size distribution with heat treatment and aging time was investigated in high strength low alloy HSLA steels. Specimen aged from 600-1000°C for 2 hours were examined using NBSR SANS facility with a linear position sensitive detector, after saturation of the magnetic

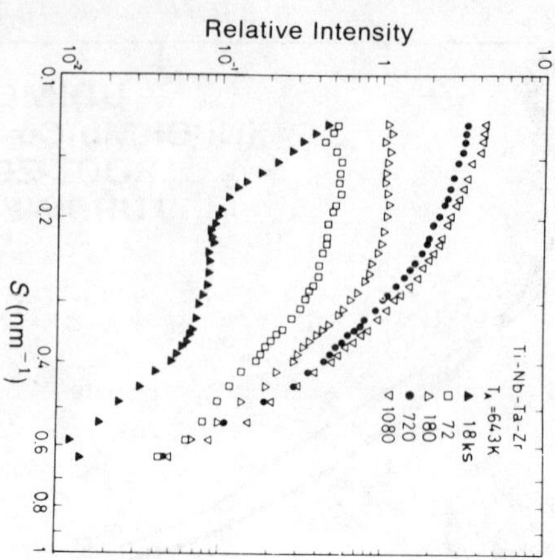

Figure 10. Coarsening of the α-phase precipitates in a superconducting wire (64).

domains. The scattering profiles processed by the usual data analysis methods. The Guinier plots (Fig. 11) as well as the calculated size distribution (Fig. 12) agreed well with results obtained using electron microscopy.

Another application of SANS to precipitation problem in dense systems at NRL is also worth mentioning. In this study the coarsening of the δ' phase transformation in an Al-Li system was studied using SANS. Preliminary SANS profiles were obtained for a few specimens, from which an estimate of the range of temperature, and annealing twins for a more extensive SANS study has been made (66). The distinguishing feature of these profiles compared to the beta-III Ti is the presence of a large forward scattering component, with the interference peak appearing as a "shoulder" on the continuous scattering intensity (Fig. 13). In contrast, the diffraction-like maxima in Ti SANS profiles were apparently related to some type of ordering among the precipitates without a true "forward scattering" from the matrix itself. Because of the presence of interparticle effect size distribution calculations on this data are difficult, and require elaborate schemes for interference term corrections.

Voids and Cavities

Voids and cavities encountered in a large number of metals, alloys and non-metallic materials are the result of many processing and thermomechanical parameters such as superplastic deformation, creep, fatigue, powder compaction, extrusion, hot isostatic pressurization (HIP),

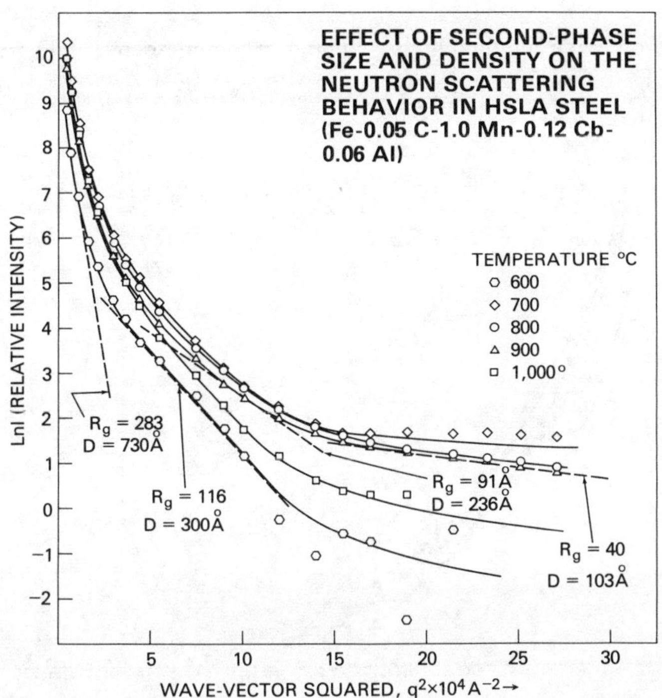

Figure 11. Guinier plot of SANS data from HSLA steels, showing the effect of second phase size and density on the neutron scattering (65).

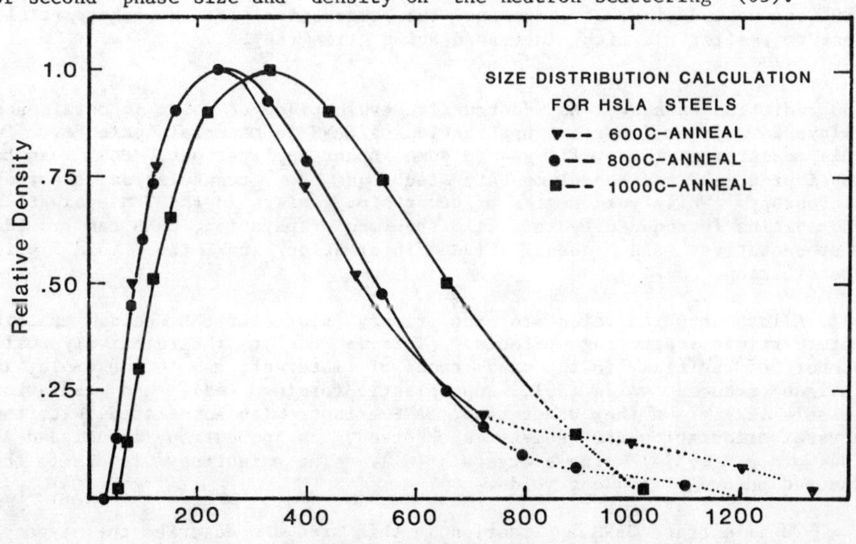

Figure 12. Size distribution calculation for three HSLA specimens for aging temperatures indicated (65).

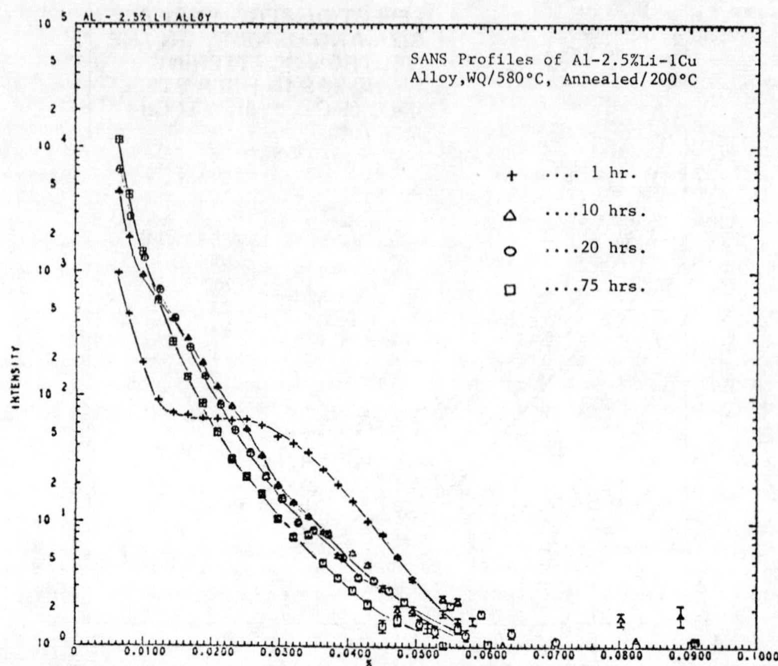

Figure 13. SANS profiles of (Al-2.5%Li-1Cu) alloy water quenched at 580°C, annealed at 200°C, showing the progression of the interparticle peak to smaller q's with increased aging time (66).

and radiation damage. Non-destructive evaluation of voids in metals and alloys is another important application of SANS in materials science. In this measurement, as well as in some other applications, SANS can be considered as the complementary technique to transmission electron microscopy. While some degree of uncertainty exists in the TEM evaluation of cavities introduced by thin film specimen preparation, SANS can provide representative and quantitative information on the void size distribution.

Alloys in which voids are the primary source of SAS seldom exhibit interparticle scattering effects. This is due to the relatively small number of cavities in the many cases of interest, as for example, in fatigue induced voids (36), superplastic forming (65), and radiation vessels (22). On the other hand, SANS anisotropies associated with the crystal orientation are sometimes observed, as in neutron irradiated Al (34) and in β-NiAl single crystals (41). The anisotropy is due to the faceted nature of induced voids.

To illustrate SANS application in this area and describe the type of information that can be obtained, an application of SANS in the investigation of fatigue induced grain boundary cavitation is discussed in some detail below.

Many metals and alloys exhibit the phenomenon of grain boundary cavitation (growth of voids at the grain boundaries) when deformed at high temperatures. The nucleation and growth of these voids are still not clearly understood. Such voids can act as scattering entities for neutrons. SANS is found to be highly sensitive for characterizing voids as void volume fraction of 10^{-6} or less can be determined (compared to about 10^{-4} from density measurements). Using SANS, Weertman and coworkers (36) have measured void nucleation rates, growth rates, void size distribution etc. which develop during creep or fatigue SANS measurements were made on high purity (99.999%) copper specimens fatigued for various times in fully reversed loading with a stress amplitude of 34 M Pa and a frequency of 17 cycles/sec. in an atmosphere of argon. The temperatures were varied from 405°C to 567°C. The SANS measurements were carried out on the D-11A instrument at the ILL. The wavelength used was 0.697 nm and q range was $0.032 < q \leq 1.905$ nm^{-1}, i.e., over a range sufficiently broad to cover both the Guinier region and Porod region. High voltage electron microcopy was used to show that the voids were roughly spherical in shape and this information was used to analyze the intensity profiles. It can be shown that the integrated intensity I is directly proportional to the total fractional particle volume $\Delta V/V$ of the sample, i.e.,

$$I = 8\pi^3(\Delta\rho)^2 \frac{\Delta V}{V} \qquad (27)$$

where $\Delta\rho$ is the difference in scattering length density between the matrix and the voids. Figure 14 shows the fractional particle volume ($\Delta V/V$) as a function of fatiguing time for three different temperatures. The results plotted give a good indication of the sensitivity of the technique.

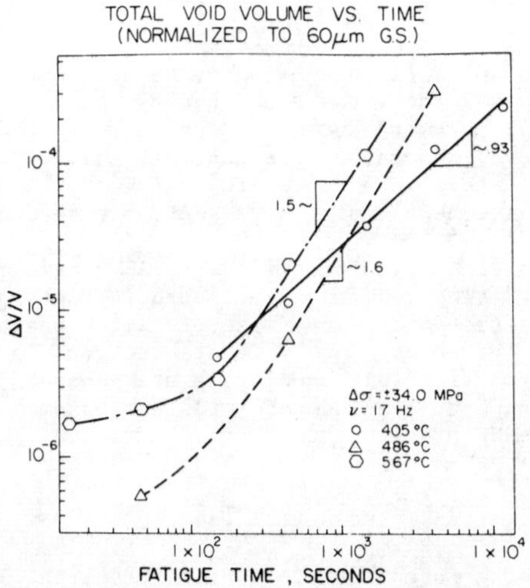

Figure 14. Total void volume vs time for fatigued copper in fully reversed loading with a stress amplitude of 34.0 MPa (39).

The $\Delta V/V$ measured is the total void volume for a given time of fatiguing. This includes newly created voids and voids that grow during the fatiguing time. SANS is able to separate the two factors in the following fashion. Since the intensity profile is known over the whole range of q one can obtain a distribution of particle sizes for each fatiguing time and temperature using standard numerical inversion schemes. This is plotted in Fig. 15. The total number of voids then can be determined by integrating the area under the size distribution curves. By repeating these measurements for various time of fatiguing, one can then obtain void nucleation rates as well as individual void growth rates. A detailed understanding of the void nucleation and growth is thus made possible by SANS. Plots of log individual void growth rate vs log time were found to be linear in agreement with theory. However, the slopes of these plots were not in such good agreement.

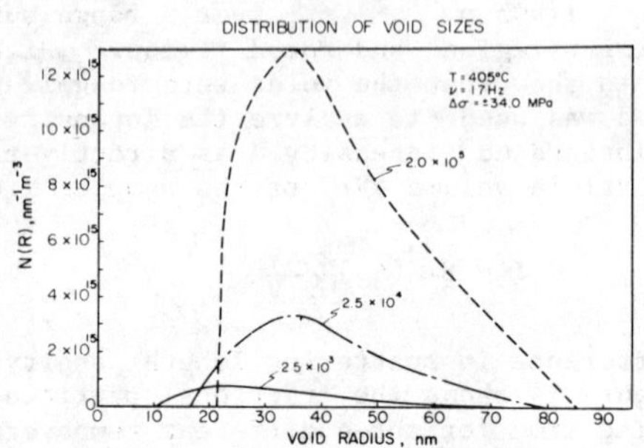

Figure 15. Void size distribution for fatigued copper of Figure 14 (39).

Some other void measurements using SANS are now briefly reviewed. Hendricks et al examined the voids generated by neutron irradiation in Al single crystals.(35) Their detector system consisted of three small detectors placed above and below the central beam. From the observed anisotropic pattern, they were able to calculate a void shape of a trunicated octahedra; somewhat similar results are obtained from TEM.

Void formation in stoichiometric β'-NiAl single crystal has been studied by Epperson. The resulting anisotropic pattern again suggested the presence of faceted voids, the size of which was calculated by the Guinier method to be about 250Å. Fast neutron irradiated GaAs was examined by Gupta et al (40) showing a star-shaped pattern of SANS anisotropy. The shape of the induced voids was assumed to be ellipsoidal and a calculated SANS profile showed good agreement with the observed data.

Industrial Use

SANS is now being used for industrial research. The Fiat Research Center has been developing SANS for industrial applications, chiefly to monitor thermal processing of complex alloys. Details of this poineering research are given in ref 20. The applications include, the study of effect of tempering temperature and austensite content on line broadening, in a complex steel, precipitates in superalloys, microvoids in Si_3N_4 during thermal fatigue, etc..

Conclusions

The examples of SANS applications discussed above demonstrate the sensitivity and versatility of the technique for detecting and characterizing defects. As is clear from Table II, the type of problems that can be characterized are fairly large, ranging from the fundamental studies of phase transformatiohns to thermal processing of complex industrial alloys. Similarly, the variety of materials studied are large and include metals, alloys, semiconductors, superconductors, glasses, and magnetic materials. SANS has thus proved to be a powerful new tool available to material scientists. Its applications to materials research and its uses as a NDE technique is expected to grow substantially in the near future.

References

1. K. Ibel, "The Neutron Small-Angle Camera D11 at the High-Flux Reactor, Grenoble," J. Appl. Cryst. 9 (1976) pp. 296-309.

2. W. C. Koehler and R. W. Hendricks, "The United States National Small-Angle Neutron Scattering Facility," J. Appl. Phys. 50 (1979) p. 1951.

3. H. R. Child and S. Spooner, "A New Small-Angle Neutron Scattering (SANS) Instrument at ORNL Using a Position Sensitive Area Detector," J. Appl. Cryst. 13 (1980) pp. 259-264.

4. C. Glinka, "Neutron Scattering," American Institute of Physics Conf. Proc. #89 (1981) pp. 395-397.

5. P. Pizzi and H. Walther, "A Cold Neutron Small-Angle Scattering (SAS) Device for Technological Applications," J. Appl. Cryst. 7 (1974) pp. 230-232.

6. D. F. R. Mildner, R. Berliner, O. A. Pringle, and J. S. King, "The Small-Angle Neutron Scattering Spectrometer at the University of Missouri Research Reactor," J. Appl. Cryst. 14 (1981) pp. 370-382.

7. C. Hofmeyr, R. M. Mayer, and D. L. Tillwick, "The Small-Angle Neutron Scattering Facility at the SAFARI-1 Reactor, J. Appl. Cryst. 12 (1979) pp. 192-200.

8. A. D. Hardy, K. D. Rouse, and M. W. Thomas, "Small-Angle Neutron Scattering Experiments, Part II, Use of the Harwell Small-Angle Scattering Spectrometer," Report No. ND-R-467(s) Part I and II. Issued by UKAEA RISLEY LANC. England, April 1981 (Available from HMSO, London, UK).

9. J. Schelten, "Die Kleinwinkelstreuanlage am Gekrummten Neutronenleiter der Katten Neutronenquelle am Reaktor FRJ-2 (The Small Angle Scattering Facility at the Bend Neutron Guide of the Cold Neutron Source at the Reactor FRJ-2)," Kerntechnik 14 (1972) pp. 86-88.

10. P. Debye and A. Bueche, "Scattering by an Inhomogeneous Solid," J. Appl. Phys. 20 (1949) pp. 518-525.

11. A. Guinier and G. Fournet, "Small-Angle Scattering for X-rays," (translated by C. B. Walker), John Wiley, New York, (1955).

12. G. Porod, "Die Rontgenkleinwinkelstreung von Dichtgepackten Kolloiden System II," Kolloid Z. 125 (1952) pp. 51-57, 109-122.

13. P. Debye, H. R. Anderson, and H. Brumberger, "Scattering by an Imhomogeneous Solid. II. The Correlation Function and Its Application," J. Appl. Phys. 28 (1957) pp. 679-683.

14. C. S. Pande (unpublished work).

15. G. E. Bacon, "Neutron Diffraction," Oxford Univeristy Press, London and New York (1975).

16. G. Kostorz (Ed.), "Neutron Scattering," Treatise on Material Science and Technology 15, Academic Press, New York (1979).

17. O. Glatter, "Determination of Particle-Size Distribution Functions from Small-Angle Scattering Data by Means of the Indirect Transformation Method," J. Appl. Cryst. 13 (1980) pp. 7-11.

18. C.G. Vonk, "On Two Methods for Determination of Particle Size Distribution Functions by Means of Small Angle X-ray Scattering," J. Appl. Cryst. 9 (1976) pp. 433-440.

19. H. A. Mook, "Neutron Small-Angle Scattering Investigation of Voids in Irradiated Materials," J. Appl. Phys. 45 (1974) pp. 43-46.

20. H. Walter and A. Pizzi, "Small Angle Neutron Scattering for Non-Destructive Testing," in Research Techniques in Non-Destructive Testing IV, edited by R. S. Sharp, Academic Press, N.Y. (1980) pp. 341-391.

21. C. Jantzen, D. Schwahn, J. Schelten, and H. Herman, "Phase Decomposition in Al_2O_3-SiO_2 Glasses," J. Appl. Cryst. 11 (1978) pp. 614-615.

22. G. Kostorz, "Neutron Small-Angle Scattering of Irradiated Aluminum Silicon Alloys," Z. Metallkunde 67 (1976) pp. 704-710.

23. G. Laslaz, P. Guyot, and G. Kostorz, "Decomposition Kinetics in Al-6.8 At% Zn," J. Physique, C7, Suppl. to No. 12, 38 (1977) pp. C7406-C7410.

24. V. Gerold, J. E. Epperson, and G. Kostorz, "On the Determination of the Metastable Miscibility Gap in Ternary Alloys from Small-Angle Measurements. II. Application to the System Al-Zn-Mg," J. Appl. Cryst. 10 (1977) pp. 28-31.

25. M. Ernst, J. Schelten, and W. Schmatz, "Neutron Small-Angle Scattering Study of the Transition from Single-to Multi-Domain Behaviour in Precipitations of Cu-1%Co," Phys. Stat. Sol. (a) 7 (1971) pp. 477-483.

26. M. Ernst, J. Schelten, and W. Schmatz, "Small-Angle Scattering of Neutrons at Single-Domain Precipitations in a Cu-1% Co Single Crystal," Phys. Stat. Sol. (a) 7 (1971) pp. 469-476.

27. A. A. Loshmanov, V. N. Sigaev, R.Ya. Khodakovskaya, N. M. Pavlushkin, and I. I. Yamzin, "Small-Angle Neutron Scattering on Silica Glasses Containing Titania," Phys. Stat. Sol. (a) 18 (1973) pp. K91-K-93.

28. A. Chamberod, M. Roth, and L. Billard, "Small Angle Neutron Scattering in Invar Alloys," J. Magn. Mag. Mat. 7 (1978) pp. 101-103.

29. S. Spooner, H. R. Child, F. H. Hsu, E. R. Vance, and J. H. Smith, "Small-Angle Neutron Scattering and Positron Annihilation in Mn-Cu Alloys," Phys. Stat. Solid. (a) 63 (1981) pp. 31-34.

30. D. Schwahn, W. Kesternich, and H. Schuster, "Evalution of Microstructural Changes in Alloy 713 LC by Neutron Small Angle Scattering and Analytical Electron Microscopy," Met. Trans. 12A (1981) pp. 155-165.

31. A. J. Leadbetter, A. F. Wright, and J. C. Knights, "Nature of the Structural Heterogeneity in SiH Films by Small Angle Neutron Scattering," Solid State Comm. 38 (1981) pp. 957-960.

32. M. Roth, and J. Zarzycki, "A Neutron Small Angle Scattering Study of SiO_2-Na_2O Glasses," J. Non-Cryst. Solids 16 (1974) pp. 93-100.

33. D. Ravaine, J. L. Souquet, and M. Roth, "The Suitability of Small-Angle Neutron Scattering for the Study of SiO_2-M_2O Glasses Outside the Immiscibility GAP," J. Non-Cryst. Solids 27 (1978) pp. 147-151.

34. H. A. Mook, "Neutron Small-Angle Scattering Investigation of Voids in Irradiated Materials," J. Appl. Phys. 45 (1974) pp. 43-46.

35. R. W. Hendricks, J. Schelten, and W. Schmatz "Studies of Voids in Neutron-Irradiated Aluminum Single Crystals," Phil. Mag. (1978) pp. 819-836.

36. P. O. Kettunen, T. Lepisto, G. Kostorz and G. Goltz, "Voids Produced by Fatigue in Copper Single Crystals of <111> Orientation," Acta Met 29 (1981) pp. 969-972.

37. G. G. Nilsson and M. Roth, "Small Angle Neutron Scattering Investigations of Voids in Creep-Deformed Alloy 800," Material Science and Engineering 50 (1981) pp. 101-108.

38. T. Saegusa, J. R. Weertman, J. B. Cohen, and M. Roth, "Small-Angle Neutron Scattering from Pores Produced in High-Temperature Fatigue," J. Appl. Cryst. (1978) pp. 602-604.

39. R. Page, J. R. Weertman, and M. Roth, "Small Angle Neutron Scattering Study of Fatigue Induced Grain Boundary Cavities," Acta Metall. 30 (1982) pp. 1357-1366.

40. S. Gupta, E. W. J. Mitchell, R. J. Stewart, and G. Kostorz, "Anisotropy in the Diffuse Small Angle Neutron Scattering of Reactor-Irradiated Gallium Arsenide," Phil. Mag. A 37 (1978) pp. 227-243.

41. R.J.R. Miller, S. Messoloras, R. J. Stewart, J. J. Thomson, and G. Kostorz, "A Small-Angle Neutron Scattering Study of the Temperature and Stress Dependence of the Microstructure of Nimonic Alloys," J. Appl. Cryst. 11 (1978) p. 583.

42. A. P. Murani, S. Roth, P. Radhakrishna, B. D. Rainford, B. R. Coles, K. Ibel, G. Goeltz, and F. Mezei, "Small Angle Critical Neutron Scattering and the Onset of Ferromagnetism in Au-Fe Alloys," J. Phys. F: Metal Phys. 6 (1976) pp. 425-432.

43. R. Cywinski, J. G. Booth, and B. D. Rainford, "Small-Angle Neutron Scattering from Superparamagnetic Assemblies in Binary CoGa Alloys," J. Phys. F: Metal Phys. 7 (1977) p. 2567.

44. B. Boucher, A. Lienard, J. P. Rebouillat, and J. Schweizer, "Neutron Study of the Magnetic Correlations in Amorphous $ErCo_2$: II. Evidence for Two Different Regimes in Small-Angle Neutron Scattering," J. Phys. F: Metal Phys. 9 (1979) pp. 1433-1440.

45. S. Komura, G. Lippmann, and W. Schmatz, "Temperature Variation of the Magnetic Cluster Structures in an Iron-Nickel Invar Alloy," J. Magn. Mag. Mat. 5 (1977) pp. 123-128.

46. Z. Wang, M. Fong, X. Shie, M. Roth, and Z. Zhi-You, "Magnetic Measurements and Small-angle Neutron Scattering (SANS) Study of Some Amorphous Fe-Ni-Mo-B Alloys," J. Magn. Mag. Mat. 28 (1982) pp. 143-148.

47. J. W. Lynn, A. Raggazoni, R. Pynn, and J. Joffrin, "Observation of Long Range Magnetic Order in the Reentrant Superconductor $HoMo_6S_8$," J. Physique -Letters 42 (1981) pp. L45-L49.

48. V.I. Goman'kov, B. N. Mokhov, and, E. I. Mal'tsev, "Ferromagnetic Regions of Polarization in an Antiferromagnetic Matrix in the System $Fe_{65}(Ni_{1-x}Mn_x)_{35}$," Pis'ma Zh. Eksp. Teor. Fiz. 23 97-100 (1976) pp. 83-86.

49. G. E. Fish, and H. R. Child, "Studies of Chemical Homogeneity and Magnetic Domain Walls in Fe-Based Metallic Glasses using Small-Angle Neutron Scattering," J. Appl. Phys. 52 (1981) pp. 1880-1882.

50. U. Denkhaus, J. Schelten, and W. Schmatz, "Magnetic Propeties in Partially Ordered Ni_3Mn", J. Appl. Cryst. 7 (1974) p. 232.

51. S. M. Shapiro, G. Shirane, B. H. Verbeek, G. J. Nieuwenhuys, and J. A. Mydosh, "Neutron Scattering Studies of the Reentrant Spin-Glass System PdFeMn," Solid State Comm. 36 (1981) pp. 167-170.

52. B. Boucher, "Neutron and X-Ray Small Angle Scattering (S.A.S.) Study of the Amorphous Alloy $Tb_{.25}Cu_{.75}$," J. Physique, Suppl to No. 8, 41 (1980) pp. C8-135-C140.

53. M-C. Bellissent-Funel, M. Roth, and P. Desre, "Small Angle Neutron Scattering on Liquid Ag-Ge Alloys," J. Phys. F: Metal Phys. 9 (1979) pp. 987-998.

54. T. Takahashi, H. Tomimitsu, Y. Ushigami, S. Kikuta, and K. Doi, "The Very-Small Angle Neutron Scattering from Neutron-Irradiated Amorphous Silica," Jap. J. Appl. Phys. 20 (1981) pp. L837-L839.

55. M. Roth, "The Small-Angle Scattering of Neutrons by Surface Imperfections," J. Appl. Cryst. 10 (1977) pp. 172-176.

56. J. R. Weertman, "Identification by Small Angle Neutron Scattering of Microstructural Changes in Metals and Alloys," Non-Destructive Evaluation, Microstructural Characterization and Reliability Strategies, edited by O. Buck and S. M. Wolf, Metallurgical Society of AIME, N. Y. (1981) pp. 147-168.

57. V. Gerold and G. Kostorz, "Small-Angle Scattering Applications to Materials Science," J. Appl. Cryst. 11 (1978) pp. 376-404.

58. M. Fatemi, C. S. Pande and H. R. Child," Detection of Several Omega Phase Morphologies in Beta-III Titanium by Small Angle Neutron Scattering and Transmission Electron Microscopy," Phil. Mag. (1983) in press.

59. S. K. Sikka, Y. K. Vohra, and R. Chidambaran, "Omega Phase in Materials," Prog. in Mat. Sci. 27 (1982) pp. 245-310.

60. B. S. Hickman, "Precipitation of the Omega Phase in Titanium-Vanadium Alloys," J. Inst. Met. 96 (1968) pp. 330-337.

61. J. M. Raynal, J. Schelten, and W. Schmatz, J. Appl. Cryst. 4 (1971) p. 511.

62. S. A. Messaloras, Ph.D. thesis, Univ. of Reading, U.K.

63. A. Boeuf, R. Coppola, R. F. Zabonaydi, S. Melone, S. Maggi and P. Puliti, "Small Angle Neutron Scattering Investigation of M23C6 Precipitation in Al51 304 Stainless Steel," 14 (1981) pp. 337-344.

64. K. Osamura, H. Tsunekawa, Y. Murakami, M. Ono, H. Yoshida, S. Okamoto, Y. Monju, and T. Fukuzuka, J. Appl. Cryst. 15 (1982) p. 611.

65. M. Fatemi and B. B. Rath, "Non-Destructive Evaluation of Defects in Structural Materials Using Long Wavelength Neutrons," NRL Memorandum Report 4226, May 1980.

66. M. Fatemi and B. B. Rath, "SANS Study of Coarsening of δ' Precipitates in an Al-Li Alloy," (1981) unpublished.

Subject Index

Author Index